*The Best American Science
and Nature Writing 2018*

The Best American Science and Nature Writing™ 2018

Edited and with an Introduction
by Sam Kean

Tim Folger, Series Editor

A Mariner Original

HOUGHTON MIFFLIN HARCOURT

BOSTON • NEW YORK 2018

hmhco.com

ISBN 978-1-328-98780-8 (print) ISSN 1530-1508 (print)
ISBN 978-1-328-99019-8 (ebook) ISSN 2573-475X (ebook)

Printed in the United States of America
DOC 10 9 8 7 6 5 4 3 2
4500747222

Contents

Contents

Foreword

EARLY IN 2017, for some strange reason, George Orwell's *Nineteen Eighty-Four* suddenly jumped to the top of Amazon's bestseller list. Orwell laced his dystopian novel with Newspeak, the language of Oceania, one of the story's perpetually warring states. Here's a short sampler of its Big Brother–approved vocabulary: *minipax* — the Ministry of Peace (Oceania's war department, not to be confused with our Department of Defense); *prolefeed*—mindless mass entertainment; *malquoted*—what today's authoritarians would call fake news. And then there's *blackwhite*, a synecdoche for all the perversions of Newspeak: to believe that black is white.

Our own leaders have given us "enhanced interrogation," "collateral damage," "clean coal," and so many more. Whatever else they might be, when it comes to contorted Orwellian syntax, they are true artists. To honor those who tirelessly seek to protect us from the dangers of clear, informed English, why don't we create a new award, the Quack, inspired by the Newspeak word *duckspeak:* communication unsullied by reason. Or, as Orwell himself defined it, "to make articulate speech issue from the larynx without involving the higher brain centres at all."

First nominee: Kathleen Hartnett White, shrewdly chosen by the president to head the Council on Environmental Quality. Ms. White-is-Black rejects the overwhelming scientific evidence that the release of carbon dioxide into the atmosphere from the burning of fossil fuels is causing catastrophic climate change. Not true, she assures us. "Carbon dioxide," she said in 2016, "is the gas of life on this planet." In an unprecedented concession to reality, the president withdrew her nomination in February. Thankfully, she's still eligible for a Quack.

And who's that, doddering down the blackwhite carpet on his way to the Quack Awards ceremony? Could it be? It is! Lamar Smith—the anti-science representative from Texas who now chairs—what else?—the House Committee on Science, Space, and Technology. During congressional testimony on climate change, he once said that the journal *Science*—one of the world's most prestigious scientific publications—"is not known as an objective magazine." As the great physicist Wolfgang Pauli once said of a colleague, Smith's arguments are so bad they're not even wrong. To be fair, he's not opposed to *all* science. He is a fervid proponent of space exploration and evidently heeds what scientists tell him about Mars and other worlds. Their warnings about threats to his home planet? Not so much. Smith would like to see a crewed mission to Mars by 2021. Maybe taxpayers could fund his fare?

It's not easy picking a winner from such a competitive and crowded field. We haven't even mentioned Energy Secretary Rick Perry, who during a presidential debate couldn't name the department he now leads. He once offered this answer to a fourth grader who asked him about Earth's age: "You know what? I don't have any idea. I know it's pretty old, so it goes back a long, long way. I'm not sure anybody actually knows completely and absolutely how long, how old the Earth is." For Perry, who describes himself as "a firm believer in intelligent design as a matter of faith and *intellect*" (emphasis added), "pretty old" apparently means 6,000 years. He's only off by a factor of 756,000 or so. As for "how long" the Earth is, well, maybe he thinks our planet is rectangular? We mustn't judge him too harshly, though. In his defense, he has said, "I am not a scientist." Good thing he made that clear.

With so many highly qualified contenders, it's tempting to award a collective Quack. But hey, this isn't a socialist country; there has to be one winner—and a slew of losers. And the clear victor is . . . Scott Pruitt, the head of the Environmental Protection Agency, the very agency he sued multiple times while serving as Oklahoma's attorney general! He is proving to be a fierce guardian of the environment, if by "environment" you mean the lands owned, leased, or otherwise coveted by the country's largest oil, gas, and mining companies.

Pruitt's leadership has been visionary—or hallucinatory, depending. He has barred any scientist who has managed to win highly competitive government funding from advising the EPA. Letting our best researchers have input on environmental issues would be an egregious conflict of interest in Pruitt's *blackwhite* "reasoning." Far better to seek wise, impartial counsel from the industries the EPA is chartered to monitor.

So it's only fitting that we honor Mr. Pruitt with the first Quack Award, as well as an entire article in this collection devoted to his disastrous tenure at the EPA. If anyone doubts that Pruitt has earned this honor, consider the opening words of Rachel Leven's "A Behind-the-Scenes Look at Scott Pruitt's Dysfunctional EPA": "Environmental Protection Agency administrator Scott Pruitt doesn't hide his contempt for how the agency has been run." From there her story becomes ever more disturbing and heartbreaking.

Maybe Pruitt is doing us all a favor, getting the government out of our lives and forcing each of us to take responsibility for the state of the world? If, one by one, we all make the right choices—if we recycle more, drive less, eat organic—can we avoid the threats of climate change without the need for organized governmental intervention? Such lifestyle changes might soothe our first-world consciences, but as David Roberts shows in "Wealthier People Produce More Carbon Pollution—Even the 'Green' Ones," well-meaning individual actions are no substitute for far-reaching national and international policy initiatives.

Fortunately, policy trends that could possibly save us are already discernible, despite the criminal efforts of Pruitt and his fellow Quacks to halt or reverse them. In this anthology, a promising new science journalist argues convincingly that reducing our civilization's carbon emissions will not necessarily hinder global economic growth. From 2008 through 2015, the author states, the U.S. economy grew by 10 percent while the amount of energy consumed per dollar "fell by almost 11 percent." For more insights from this perspicacious writer, dig into "The Irreversible Momentum of Clean Energy." No doubt we'll see more fine articles from Barack Obama in the years ahead. President Obama's article was published in *Science*, which, despite Lamar Smith's objections, happens to be one of the most credible and prescient sources of information in the world.

With climate change threatening to unmoor our global civilization, we'll certainly need to look ahead, to imagine how our decisions today might prevent or accelerate our slide to ruin. But we can also learn by looking backward, to the origins of the world's very first civilizations. How did they start? Where did they start? What drove our ancestors to give up their old hunter-gatherer ways? In "The Case Against Civilization," John Lanchester proposes that somewhere back in prehistory we took a collective wrong turn when we began to value the accumulation of wealth above all other human endeavors. Is it too late for us, Lanchester asks, to change our direction? Quoting John Maynard Keynes, he wonders if one day we'll come to see the pursuit of riches above all else as a "disgusting morbidity."

Think of this collection as a perfect antidote to quackery, and to the *prolefeed* that Orwell warned us about. Sam Kean, this year's guest editor, has selected stories for you that Big Brother would have condemned as fostering *ownlife:* an appreciation of the individual and the value of solitude and reflection. Stories like Elena Passarello's shimmering "Arabella," a meditation on space exploration, almost elude classification and might never garner the "likes" or tweets they deserve. That is equally true of Caitlin Kuehn's "Of Mothers and Monkeys," an unforgettable essay on illness, death, and cross-species empathy. Sam, who has devised a provocative presentation for these wonderful articles, will take it from here.

Next year, *The Best American Science and Nature Writing* will have a new series editor, Jaime Green. Writers, editors, and readers can nominate articles for the anthology by following the submission guidelines on her website: https://www.jaimegreen.net/basn. After 17 years of helping to edit this anthology, I'll have the distinct pleasure of joining the ranks of its more casual readers. (I'm looking forward to enjoying the book without having done a bit of the legwork!) As always, I'm grateful to Naomi Gibbs at Houghton Mifflin Harcourt and many others there without whom this series wouldn't exist. And I'd like to put a bit of Newspeak to good use to thank my beauteous wife, Anne Nolan. Like this anthology, she is *doubleplusgood* ("the best").

TIM FOLGER

Introduction

RIGHT AROUND THE time I moved back home after college (which was humiliating enough), my parents started feeding stray cats in our backyard. They'd buy these huge tins of Walmart cat food, and when you opened them up and turned them over, nothing happened at first. Only after several seconds would the hockey puck of horsemeat inside start quivering. It would then make this sucking noise as it plopped out, followed by glistening strands of gravy-mucus. It smelled awful—probably was offal. But because this was wintertime in South Dakota, the strays got pretty desperate for even this food. While lying in bed at night sometimes, I'd hear shrieks and howls from the backyard. I'd peek through my bedroom curtains and see this Tasmanian-devil tornado of fur whirling around out there. Not uncommonly, there'd be streaks of blood on the concrete the next morning.

Clearly, it made sense to take some of these cats into our home as pets. "They deserve a good life, too," my mother insisted. The first cat was a long-haired Maine Coon tabby we named Ignatius. Then another tabby we named Madeline. And—life's little blessings—Madeleine was pregnant, so out popped Oliver and Cassandra.

Because they were fairly tame indoors, once they had food and warmth, I mostly ignored the cats—except for one thing. I'd majored in science in college but had since slacked off, letting that part of my brain grow flabby. I wanted to get it back in shape, so I picked up a few meaty science books, including a long tome on animal behavior. It was pretty dense—I couldn't read more than a few pages each night, and it took me months to finish some chapters. I certainly didn't understand all of it either. But one theme stuck in my mind: that animal behavior is largely programmed. Sometimes genes drive their behavior, sometimes hormones or neural circuits, sometimes environmental cues. But whatever the cause, animals act and react in highly predictable ways, the book argued. In some cases you could even describe their behavior with precise equations.

Now, I knew this was controversial stuff. The author was an ant expert, and it made sense to think of ants as these little chemical zombies running around with no free will. But then he started working his way through the animal kingdom, applying the same ideas to fish, reptiles, birds—and mammals. Like cats. And that's when the book started to get a little spooky. Because one example it mentioned was that overcrowding can drastically change an animal's behavior. Animals get much more territorial if there are too many other creatures around, and even docile individuals can turn aggressive and nasty.

So why was this spooky? Because right around the time I read this, my mom, bless her heart, took in a fifth cat, Elliott. And it was like a phase shift occurred—the Kean cat population had vaulted beyond some critical barrier.

Ignatius started pouncing on the other cats from windowsills. Cassandra suffered gashes on her face and started tearing up furniture with her claws. Worst of all, Oliver and Elliott had a literal pissing match—spraying curtains and sofas to mark territory. The book had predicted that things like this would happen, and seeing the drama unfold like that, right in our living room, was uncanny. I felt like I'd stumbled into some sort of cabbalistic text about the animal kingdom, a shortcut to all its secrets. This was powerful stuff.

Still, however spooky the scene was, at the end of the day these were just cats. Beasts. *Of course* something like overcrowding could override their basic decency.

Then, a few months later, my brother also moved back home after college. Now we had four adult *Homo sapiens* living in one house. And honestly, we had a nice honeymoon of family togetherness at first—my family really can be great. But eventually that long South Dakota winter descended again and cooped us all indoors.

The changes were subtle at first. It seemed like someone was always stomping up and down the stairs just to annoy me, or knocking on the bathroom door when I needed some privacy. And day by day I began to feel more—I guess the word is *territorial*. I would hole up in my bedroom instead of interacting with anyone, and would squirrel all my groceries away at the back of the cabinet so no one else could eat them. I didn't do any of this consciously, mind you. It just felt natural.

Not long afterward, the fights started. My mother would say something completely unreasonable like, "So, do you have any plans this weekend?" I didn't, obviously, so I'd jab back. Things would escalate, and when I finally went too far, she would actually grunt at me. Her neck would swell up, her face would flush red, and she'd unleash this guttural scream—*grrrraauuugh*. I'd get so sputtering angry that I couldn't speak either. I'd clench my fists and bare my lower canines instead. I had no idea, but these were classic territorial displays.

For his part, my brother said one day, "I have got to get out of this zoo." So he joined a gym and started lifting, packing 30 extra pounds of muscle onto his shoulders and legs. He basically became the alpha male, the silverback gorilla of the house.

Meanwhile, there I was, still completely oblivious to what was really going on, even as I worked my way through the book's final section, on primate behavior. One thing this section mentioned was that primates have more complicated reactions to overcrowding than other species; they'll sometimes withdraw from social contact, for instance. Well, funny thing: my father would often run these useless errands to his office at night, just to leave the house. Weirdly, too, my brother and I weren't going out much, and I'd more or less become nocturnal, just to find some peace and quiet. We also grew these hideously patchy neck beards, and we probably weren't washing our hair as often as young men should. We'd basically stopped grooming ourselves.

In fact, I was examining my new beard in the mirror one night, fuming over some now-forgotten slight, when I realized something: that I'd reverted to a savage state, just like the cats had. The book had nailed *my* behavior too.

This was a real punch in the gut. We all want to think of ourselves as independent and autonomous, as fully in charge of our behavior and emotions—but you can't fool biology. Even worse, all I could think to do at this point was keep reading ahead in the book and see what would happen to me now. It was like finding a biography of yourself while you're still alive—you can't help but flip to the end.

My epiphany had two effects on me. First, it convinced me that I really, really needed to move out before we went full feral and started flinging ungodly things at each other. Second, it helped re-route my life. If it wasn't obvious, I'd been drifting for several years at that point, unsure what to do with myself. Hoping to maximize my unemployment prospects, I'd gotten degrees in physics and English literature in college, then decided to write. But as stupid as it sounds, it had never really occurred to me to write *about science*. That book changed my outlook, shifted my *weltanschauung*. Even more than that, it shook up my view of what science and science writing could be. Before this, I'd always approached science like a logic game or riddle. I'd solve the problem at hand and feel a little ping of pleasure when I got the right answer; it pleased me the same way a perfectly filled-out crossword puzzle did, how neat and tidy everything looked.

This experience of science was different—very personal, very potent. I wasn't even happy about it necessarily; the insights definitely made me uncomfortable. But I probably needed some discomfort then, and like Kant reading Hume, it did awaken me from my "dogmatic slumbers." I wasn't sure what exactly I wanted to write about yet, or how to start, but whatever reaction this book provoked, I wanted more of it.

So that summer I applied for a science-writing internship in a faraway city, St. Louis, where I knew no one. And as soon as I announced plans to leave, of course, that honeymoon of family togetherness descended again. On the day I left, my parents both hugged me goodbye, hard, and as I pulled out of our driveway, I actually got choked up. Because for all that I'd hated being there, another big theme of the book, besides the aggression stuff, was that family is deeply rooted inside us. That's as biological as anything.

At the same time, I took comfort in something else the book told me: that in many species of primates, individuals disperse at a certain age. That is, they leave their families and join another troop somewhere. It's never easy for them—some keep looking back every few yards, afraid to abandon all they knew. But it's the law of the wild, and it was high time for this allegedly adult ape to strike out on his own.

Four books and probably a million words later, here I am, still chasing that dragon. I feel immensely lucky to write about science for a living, and not only because it's every bit as satisfying as I'd hoped back in my feral days. The truth is, this is one of the most exciting times in the history of science. Things aren't perfect by any means. But there are more scientists making more discoveries in more places about more things than ever before, and it's a privilege to have a front-row seat.

Perhaps not coincidentally, science writing itself has never been better either. There's always been a misconception among the public that science is Vulcan, a strictly logical enterprise. In reality it's an intensely human activity, employing the full range of both reason and emotions, of logos and pathos. And the best science writing captures all that—there's conflict and characters and drama in these tales, a real sense of craft and storytelling. (If you don't believe me, start reading this collection with Caitlin Kuehn's "Of Mothers and Monkeys" and Kenneth Brower's "The Starship or the Canoe.") It's always hard to judge greatness in its own time; you can look pretty foolish down the road. But damn the torpedoes: I think we're living in an age of great science writing.

That's why I'm thrilled to introduce the 2018 edition of *The Best American Science and Nature Writing.* Among the two dozen or so pieces here, I know you'll find stories that make you think. I also hope you'll find stories that move you and stories that make you laugh—even stories that, yes, make you uncomfortable. Because that's an important part of science too: it constantly forces us to revise our beliefs, however cherished. It doesn't give a whistle about our opinions.

I easily could have included more. To be frank, winnowing down the 100+ nominees I received was a gigantic pain in the catookus. When the individual pieces started trickling into my inbox, I immediately opened three folders on my computer: yes, no, maybe. After finishing each one I would weigh different factors and make a decision about it (or with the maybes, decide to decide later). The only problem was, the yes folder was soon bulging ominously; I couldn't bear to let anything go. Even with the maybes, I found myself thinking about them long after I'd finished reading—they wouldn't leave me alone.

In short, my filing system proved useless. I had to read half of the stories twice, a few three times, and make some excruciating decisions. So I want to thank you, science writers of the world, for two months of agony. I thoroughly enjoyed it.

I'm sure editing *BASNW* is always a challenge, but this year in particular it was agonizing. Beyond just having to leave some great pieces behind, the subject matter of many of the stories is painful. And I put off writing this introduction as long as editorially possible, because there was something I dreaded having to address.

It's no big mystery what. There was an election in the United States in late 2016, and if you're even bothering to pick up a book about science and/or writing, I can guess how you felt about the outcome. But no matter what your personal beliefs, the election fractured the country—or to be more accurate, deepened a long-standing fissure. Many people felt bitter and disappointed about the results, including (it's safe to say) the majority of scientists, probably the vast majority. Two years on from the election, the shadow of that day still looms over the scientific world, and it's been an exhausting ordeal for everyone. Those who are political wake up outraged or dispirited every day. Those who aren't political—who simply love tramping around in forests or tinkering in the lab or teaching science to children, and who would be happy to ignore politics for the rest of their lives—suddenly can't. I had a magazine editor tell me recently that she just couldn't consider science stories these day: "Politics," she said, "is sucking up all the oxygen." That struck me as incredibly depressing. Not because I missed the chance to write a story or two—that's small beer. What depressed me was the sentiment itself: that there wasn't room to cover science now (or the arts, or a dozen other things) because of a few thousand ballots on a random Tuesday in a year divisible by four. Ugh.

So that's how things stand. Science and science writing have never been stronger, never been richer. But with the country so angry, they're not getting the attention they deserve. So what, if anything, should a humble little anthology do in trying to capture this bizarre year?

I sounded out friends and colleagues about this, and (fittingly) they had divided opinions. The first approach was to ignore politics completely. Create a little oasis, they said, a place free from all that crap. Science is more dignified—it will rise above, will outlast any single man or administration, no matter how coarse or ugly. Politics breve, science larga, you might say. (And there are indeed pieces this year, like John Lanchester's "The Case Against Civilization" and Susannah Felts's "Astonish Me," that provide exactly this sort of cosmic perspective.)

Other people suggested a second approach: go all-in on politics. This is a crisis for science, they argued, in fact, a crisis for the very type of open, democratic society that makes science possible. The state of the Union is too imperiled to focus on anything else. And these people have a point. The current leaders of some of the federal agencies tasked with upholding scientific standards are in no way qualified to lead those agencies, and in a few cases are actively hostile toward the agencies' missions. Not only that but the White House seems to hold science in contempt: it still has not appointed a national science adviser to help shape policy, and its proposed travel and immigration bans are threatening to keep talented young scientists out of U.S. universities for generations. (Lest you think this is all partisan carping, check out Rachel Leven's eye-opening "A Behind-the-Scenes Look at Scott Pruitt's Dysfunctional EPA," wherein science advisers to Presidents Nixon, Reagan, and both Bushes all criticize the current administration.)

But filling a science anthology with nothing but political stories would be a mistake, I think. First of all, it would punish all the wonderful apolitical pieces that appeared this year. Moreover, when putting together an anthology, you aspire to something that will last, with stories that will be just as fresh and exciting 20 years from now as they are today. And one long political diatribe would almost certainly feel dated. Politics doesn't need to dominate every thought every day.

In the end, the selections here try to split the difference. You'll find mostly pure science pieces, pieces that celebrate the sheer wonder of the natural world. Take Ed Yong's "Tiny Jumping Spiders Can See the Moon," which (to mangle Robert Frost) begins on Twitter and ends in wisdom, combining astronomy with arachnology in a way I never dreamed possible. Or take "The Detective of Northern Oddities" by Christopher Solomon, which tracks the wonderfully eclectic career of a wildlife pathologist in rural Alaska.

Another piece, "Fantastic Beasts and How to Rank Them" by Kathryn Schulz, trucks exclusively in mythical creatures like Bigfoot and Nessie, and tries to understand why some seem more plausible to us than others. In doing so, she reveals more than you'd ever imagine about how your own mind works. I think science writers have a built-in advantage over writers who cover other topics, in that our domain has already touched nearly every human being at some point. Every child has been captivated by a wild animal, or stared up at the night sky and said *wow*. (In comparison, how many of us have ever been moved, even slightly, by a theorem in economics?) Stories like these tap into that latent wonder and show us our best scientific selves.

Other pieces here are more intentionally uncomfortable—if no less smart, hard-nosed, and well-written. Some pieces take on big, structural problems in society. Even before #MeToo, the scientific community had been rocked by a number of sexual-assault scandals, and Kayla Webley Adler's "Female Scientists Report a Horrifying Culture of Sexual Assault" uses several personal stories to evoke the hell some women endure during their careers, simply for pursuing something they love. But as serious as the human challenges are, we can't lose track of the devastating toll our activities are taking on the environment. In "Firestorm," Douglas Fox explores the astonishing science behind a forest inferno and reveals just how brittle many of our western landscapes have become. And in a more fanciful vein, "Pleistocene Park" by Ross Andersen discusses the prospects of resurrecting a long-lost grassland in Siberia, complete with lab-grown woolly mammoths, in an effort to stem destructive climate change.

Still, while there's plenty of cause for concern out there, on land and air and sea, not all the activist stories here are doom and gloom. One of the most bracing, a nonpartisan plea on "The Irreversible Momentum of Clean Energy," comes from the pen of one Barack Obama. He's still somewhat better known in fields beyond science writing, but I thought we all could use a little hope.

There's one more thing I want to mention—a sort of third way to address the state of science nowadays. We currently live in an era of Big Science. Papers on the results of large medical trials can have hundreds of authors. Genome-sequencing papers might have thousands. The peak of "hyperauthorship" so far, a paper pinning down the mass of the Higgs boson, required 24 full pages (of 33 total) simply to list all 5,154 contributors.

But scientists have never come together before and spoken with one voice quite like they did in April 2017, during the March for Science in Washington, D.C., and 600 other locations worldwide. All told, the protests drew an estimated 1.07 million men and women—some truly big-ass science. The movement even garnered cross-species support, with the live-bird "March of the Penguins for Science" at the Monterey Bay Aquarium in California. (The video is every bit as adorable as you'd think.)

To be sure, the marches were controversial in some circles, and some observers wondered whether holding them was even a good idea. After all, perhaps the greatest power of science is its air, even aura, of objectivity. There's no such thing as a truly objective individual, of course, but science as a whole—the collective work of many, many individuals—is still the best tool for getting at the truth we've ever devised. It took centuries to build up that aura, and there's a risk of losing it, of becoming just another political faction, when scientists take to the streets.

But there's no denying that the marches were unprecedented, and if this anthology really aims to capture the state of science during the past year, it couldn't ignore them. Laying aside actual research, this was arguably the biggest science story of the year.

So how to capture the marches? Obviously, we could have included a story or two about them—but *meh*. How could they hope to capture the energy of actually being there? In my own mind, I kept circling back to the *people* that day, the scientists and teachers and technicians chanting and carrying signs. And I realized that some of the best "science writing" of the year appeared right there, on those signs. We all knew scientists were smart, but who knew they could be so pithy and witty and cutting as well?

It seemed appropriate, then, to highlight that wit and wisdom here. Unlike most previous editions of *BASNW*, the stories here aren't listed alphabetically by author but grouped into sections by theme. And the title of each section comes directly from a sign at one of the marches. Space stories, for instance, fall under "It's Not Rocket Science . . . (Actually, Some of It Is)." Profiles of different scientists are gathered under "So Bad, Even the Introverts Are Here." There is "At the Start of Every Disaster Movie Is a Scientist Being Ignored" for stories on environmental science, and "What

Do We Want? Evidence-Based Science. When Do We Want It? After Peer Review." for stories on rethinking some current paradigms. And so on. I've never had more fun putting together a table of contents.

As I said before, science is a *human* activity above all, reflecting and refracting all of our passions and obsessions. Former Supreme Court justice Earl Warren once said, "I always turn to the sports pages first, which records people's accomplishments. The front page has nothing but man's failures." I think we've all been disillusioned about athletes at this point, but that still holds true in the main for science—it's largely a record of our accomplishments, however imperfect so far. And when times get bleak, science can help us find direction, too. In "Greetings, E.T. (Please Don't Murder Us.)," Steven Johnson discusses the promise and peril of making contact with an alien society, and he writes at one point, "Thinking hard about what kinds of civilization we might be able to talk to ends up making us think even harder about what kind of civilization we want to be ourselves." We could all stand to think about that right now, and I hope these stories can help us along that path—both in making us uncomfortable with the limitations of now and in pointing to where we can get to in the future.

Sam Kean

PART I

"We Are Not Just Resistors,
We Are Transformers"

Transformative Science

ROSS ANDERSEN

Pleistocene Park

FROM *The Atlantic*

NIKITA ZIMOV'S NICKNAME for the vehicle seemed odd at first. It didn't look like a baby mammoth. It looked like a small tank, with armored wheels and a pit bull's center of gravity. Only after he smashed us into the first tree did the connection become clear.

We were driving through a remote forest in Eastern Siberia, just north of the Arctic Circle, when it happened. The summer thaw was in full swing. The undergrowth glowed green, and the air hung heavy with mosquitoes. We had just splashed through a series of deep ponds when, without a word of warning, Nikita veered off the trail and into the trees, ramming us into the trunk of a young 20-foot larch. The wheels spun for a moment, and then surged us forward. A dry crack rang out from under the fender as the larch snapped cleanly at its base and toppled over, falling in the quiet, dignified way that trees do.

I had never seen Nikita happier. Even seated behind the wheel, he loomed tall and broad-shouldered, his brown hair cut short like a soldier's. He fixed his large ice-blue eyes on the fallen tree and grinned. I remember thinking that in another age, Nikita might have led a hunter-gatherer band in some wildland of the far north. He squeezed the accelerator, slamming us into another larch, until it too snapped and toppled over, felled by our elephantine force. We rampaged 20 yards with this same violent rhythm —churning wheels, cracking timber, silent fall—before stopping to survey the flattened strip of larches in our wake.

"In general, I like trees," Nikita said. "But here, they are against our theory."

Behind us, through the fresh gap in the forest, our destination shone in the July sun. Beyond the broken trunks and a few dark tree-lined hills stood Pleistocene Park, a 50-square-mile nature reserve of grassy plains roamed by bison, musk oxen, wild horses, and maybe, in the not-too-distant future, lab-grown woolly mammoths. Though its name winks at *Jurassic Park,* Nikita, the reserve's director, was keen to explain that it is not a tourist attraction, or even a species-resurrection project. It is, instead, a radical geo-engineering scheme.

"It will be cute to have mammoths running around here," he told me. "But I'm not doing this for them, or for any other animals. I'm not one of these crazy scientists that just wants to make the world green. I am trying to solve the larger problem of climate change. I'm doing this for humans. I've got three daughters. I'm doing it for them."

Pleistocene Park is named for the geological epoch that ended only 12,000 years ago, having begun 2.6 million years earlier. Though colloquially known as the Ice Age, the Pleistocene could easily be called the Grass Age. Even during its deepest chills, when thick, blue-veined glaciers were bearing down on the Mediterranean, huge swaths of the planet were coated in grasslands. In Beringia, the Arctic belt that stretches across Siberia, all of Alaska, and much of Canada's Yukon, these vast plains of green and gold gave rise to a new biome, a cold-weather version of the African savanna called the Mammoth Steppe. But when the Ice Age ended, many of the grasslands vanished under mysterious circumstances, along with most of the giant species with whom we once shared this Earth.

Nikita is trying to resurface Beringia with grasslands. He wants to summon the Mammoth Steppe ecosystem, complete with its extinct creatures, back from the underworld of geological layers. The park was founded in 1996, and already it has broken out of its original fences, eating its way into the surrounding tundra scrublands and small forests. If Nikita has his way, Pleistocene Park will spread across Arctic Siberia and into North America, helping to slow the thawing of the Arctic permafrost. Were that frozen underground layer to warm too quickly, it would release some of the world's most dangerous climate-change accelerants into the atmosphere, visiting catastrophe on human beings and millions of other species.

In its scope and radicalism, the idea has few peers, save perhaps the scheme to cool the Earth by seeding the atmosphere with silvery mists of sun-reflecting aerosols. Only in Siberia's empty expanse could an experiment of this scale succeed, and only if human beings learn to cooperate across centuries. This intergenerational work has already begun. It was Nikita's father, Sergey, who first developed the idea for Pleistocene Park, before ceding control of it to Nikita.

The Zimovs have a complicated relationship. The father says he had to woo the son back to the Arctic. When Nikita was young, Sergey was, by his own admission, obsessed with work. "I don't think he even paid attention to me until I was twenty," Nikita told me. Nikita went away for high school, to a prestigious science academy in Novosibirsk, Siberia's largest city. He found life there to his liking, and decided to stay for university. Sergey made the journey to Novosibirsk during Nikita's freshman year and asked him to come home. It would have been easy for Nikita to say no. He soon started dating the woman he would go on to marry. Saying yes to Sergey meant asking her to live, and raise children, in the ice fields at the top of the world. And then there was his pride. "It is difficult to dedicate your life to someone else's idea," he told me.

But Sergey was persuasive. Like many Russians, he has a poetic way of speaking. In the Arctic research community, he is famous for his ability to think across several scientific disciplines. He will spend years nurturing a big idea before previewing it for the field's luminaries. It will sound crazy at first, several of them told me. "But then you go away and you think," said Max Holmes, the deputy director of Woods Hole Research Center, in Massachusetts. "And the idea starts to makes sense, and then you can't come up with a good reason why it's wrong."

Of all the big ideas that have come spilling out of Sergey Zimov, none rouses his passions like Pleistocene Park. He once told me it would be "the largest project in human history."

As it happens, human history began in the Pleistocene. Many behaviors that distinguish us from other species emerged during that 2.6-million-year epoch, when glaciers pulsed down from the North Pole at regular intervals. In the flood myths of Noah and Gilgamesh, and in Plato's story of Atlantis, we get a clue as to what it was like when the last glaciation ended and the ice melted and the seas welled up, swallowing coasts and islands. But human culture has preserved no memory of an *oncoming* glaciation. We can only imagine what it was like to watch millennia of snow pile up into ice slabs that pushed ever southward. In the epic poems that compress generations of experience, a glaciation would have seemed like a tsunami of ice rolling down from the great white north.

One of these 10,000-year winters may have inspired our domestication of fire, that still unequaled technological leap that warmed us, warded away predators, and cooked the calorie-dense meals that nourished our growing brains. On our watch, fire evolved quickly, from a bonfire at the center of camp to industrial combustion that powers cities whose glow can be seen from space. But these fossil-fueled fires give off an exhaust, one that is pooling, invisibly, in the thin shell of air around our planet, warming its surface. And nowhere is warming faster, or with greater consequence, than the Arctic.

Every Arctic winter is an Ice Age in miniature. In late September, the sky darkens and the ice sheet atop the North Pole expands, spreading a surface freeze across the seas of the Arctic Ocean, like a cataract dilating over a blue iris. In October, the freeze hits Siberia's north coast and continues into the land, sandwiching the soil between surface snowpack and subterranean frost. When the spring sun comes, it melts the snow, but the frozen underground layer remains. Nearly a mile thick in some places, this Siberian permafrost extends through the northern tundra moonscape and well into the taiga forest that stretches, like an evergreen stripe, across Eurasia's midsection. Similar frozen layers lie beneath the surface in Alaska and the Yukon, and all are now beginning to thaw.

If this intercontinental ice block warms too quickly, its thawing will send as much greenhouse gas into the atmosphere each year as do all of America's SUVs, airliners, container ships, factories, and coal-burning plants combined. It could throw the planet's climate into a calamitous feedback loop, in which faster heating begets faster melting. The more apocalyptic climate-change scenarios will be in play. Coastal population centers could be swamped. Oceans could become more acidic. A mass extinction could rip its way up from the plankton base of the marine food chain. Megadroughts could expand deserts and send hundreds of millions of refugees across borders, triggering global war.

"Pleistocene Park is meant to slow the thawing of the permafrost," Nikita told me. The park sits in the transition zone between the Siberian tundra and the dense woods of the taiga. For decades, the Zimovs and their animals have stripped away the region's dark trees and shrubs to make way for the return of grasslands. Research suggests that these grasslands will reflect more sunlight than the forests and scrub they replace, causing the Arctic to absorb less heat. In winter, the short grass and animal-trampled snow will offer scant insulation, enabling the season's freeze to reach deeper into the Earth's crust, cooling the frozen soil beneath and locking one of the world's most dangerous carbon-dioxide lodes in a thermodynamic vault.

To test these landscape-scale cooling effects, Nikita will need to import the large herbivores of the Pleistocene. He's already begun bringing them in from far-off lands, two by two, as though filling an ark. But to grow his Ice Age lawn into a biome that stretches across continents, he needs millions more. He needs wild horses, musk oxen, reindeer, bison, and predators to corral the herbivores into herds. And, to keep the trees beaten back, he needs hundreds of thousands of resurrected woolly mammoths.

As a species, the woolly mammoth is fresh in its grave. People in Siberia still stumble on frozen mammoth remains with flesh and fur intact. Some scientists have held out hope that one of these carcasses may contain an undamaged cell suitable for cloning. But *Jurassic Park* notwithstanding, the DNA of a deceased animal decays quickly. Even if a deep freeze spares a cell the ravenous microbial swarms that follow in death's wake, a few thousand years of cosmic rays will reduce its genetic code to a jumble of unreadable fragments.

You could wander the entire Earth and not find a mammoth cell with a perfectly preserved nucleus. But you may not need one. A mammoth is merely a cold-adapted member of the elephant family. Asian elephants in zoos have been caught on camera making snowballs with their trunks. Modify the genomes of elephants like those, as nature modified their ancestors' across hundreds of thousands of years, and you can make your own mammoths.

The geneticist George Church and a team of scientists at his Harvard lab are trying to do exactly that. In early 2014, using CRISPR, the genome-editing technology, they began flying along the rails of the Asian elephant's double helix, switching in mammoth traits. They are trying to add cold-resistant hemoglobin and a full-body layer of insulating fat. They want to shrink the elephant's flapping, expressive ears so they don't freeze in the Arctic wind, and they want to coat the whole animal in luxurious fur. By October 2014, Church and his team had succeeded in editing 15 of the Asian elephant's genes. Late last year he told me he was tweaking 30 more, and he said he might need to change only 50 to do the whole job.

When I asked Beth Shapiro, the world's foremost expert in extinct species' DNA, about Church's work, she gushed. "George Church is awesome," she said. "He's on the right path, and no one has made more progress than him. But it's too early to say whether it will take only 50 genes, because it takes a lot of work to see what each of those changes is going to do to the whole animal."

Even if it takes hundreds of gene tweaks, Church won't have to make a perfect mammoth. If he can resculpt the Asian elephant so it can survive Januarys in Siberia, he can leave natural selection to do the polishing. For instance, mammoth hair was as long as 12 inches, but shorter fur will be fine for Church's purposes. Yakutian wild horses took less than 1,000 years to regrow long coats after they returned to the Arctic.

"The gene editing is the easy part," Church told me, before I left for Pleistocene Park. Assembling the edited cells into an embryo that survives to term is the real challenge, in part because surrogacy is out of the question. Asian elephants are an endangered species. Few scientists want to tinker with their reproductive processes, and no other animal's womb will do. Instead, the embryos will have to be nurtured in an engineered environment, most likely a tiny sac of uterine cells at first, and then a closet-sized tank where the fetus can grow into a fully formed, 200-pound calf.

No one has yet brought a mammal to term in an artificial environment. The mammalian mother-child bond, with its precisely timed hormone releases, is beyond the reach of current biotechnology. But scientists are getting closer with mice, whose embryos have now stayed healthy in vitro for almost half of their 20-day gestation period. Church told me he hopes he'll be manufacturing mice in a lab within five years. And though the elephant's 22-month gestation period is the longest of any mammal, Church said he hopes it will be a short hop from manufacturing mice to manufacturing mammoths.

Church has been thinking about making mammoths for some time, but he accelerated his efforts in 2013, after meeting Sergey Zimov at a de-extinction conference in Washington, D.C. Between sessions, Sergey pitched him on his plan to keep Beringia's permafrost frozen by giving it a top coat of Ice Age grassland. When he explained the mammoth's crucial role in that ecosystem, Church felt compelled to help. He told me he hopes to deliver the first woolly mammoth to Pleistocene Park within a decade.

Last summer, I traveled 72 hours, across 15 time zones, to reach Pleistocene Park. After Moscow, the towns, airports, and planes shrunk with every flight. The last leg flew out of Yakutsk, a gray city in Russia's far east, whose name has, like Siberia's, become shorthand for exile. The small dual-prop plane flew northeast for four hours, carrying about a dozen passengers seated on blue-felt seats with the structural integrity of folding chairs. Most were indigenous people from Northeast Siberia. Some brought goods from warmer climes, including crops that can't grow atop the permafrost. One woman held a bucket of grapes between her knees.

We landed in Cherskiy, a dying gold-mining town that sits on the Kolyma River, a 1,323-mile vein of meltwater, the largest of several that gush out of northeastern Russia and into the East Siberian Sea. Stalin built a string of gulags along the Kolyma and packed them with prisoners, who were made to work in the local mines. Solzhenitsyn called the Kolyma the gulag system's "pole of cold and cruelty." The region retains its geopolitical cachet today, on account of its proximity to the Arctic Ocean's vast undersea oil reserves.

Cherskiy's airstrip is one of the world's most remote. Before it became a Cold War stronghold, it was a jumping-off point for expeditions to the North Pole. You need special government permission to fly into Cherskiy. Our plane had just rolled to a halt on the runway's patchy asphalt when Russian soldiers in fatigues boarded and bounded up to the first row of the cabin, where I was sitting with Grant Slater, an American filmmaker who had come with me to shoot footage of Pleistocene Park. I'd secured the required permission, but Slater was a late addition to the trip, and his paperwork had not come in on time.

Nikita Zimov, who met us at the airport, had foreseen these difficulties. Thanks to his lobbying, the soldiers agreed to let Slater
through with only 30 minutes of questioning at the local military
base. The soldiers wanted to know whether he had ever been to
Syria and, more to the point, whether he was an American spy. "It
is good to be a big man in a small town," Nikita told us as we left
the base.

Nikita runs the Northeast Science Station, an Arctic research
outpost near Cherskiy, which supports a range of science projects
along the Kolyma River, including Pleistocene Park. The station
and the park are both funded with a mix of grants from the European Union and America's National Science Foundation. Nikita's
family makes the 2,500-mile journey from Novosibirsk to the station every May. In the months that follow, they are joined by a
rotating group of more than 60 scientists from around the world.
When the sky darkens in the fall, the scientists depart, followed by
Nikita's family and finally Nikita himself, who hands the keys to a
small team of winter rangers.

We arrived at the station just before dinner. It was a modest
place, consisting of 11 hacked-together structures, a mix of laboratories and houses overlooking a tributary of the Kolyma. Station
life revolved around a central building topped by a giant satellite
dish that once beamed propaganda to this remote region of the
Soviet empire.

I'd barely stepped through the door that first night when Nikita offered me a beer. "Americans love IPAs," he said, handing
me a 32-ounce bottle. He led us into the station's dining hall, a
warmly lit, cavernous room directly underneath the satellite dish.
During dinner, one of the scientists told me that the Northeast Science Station ranks second among Arctic outposts as a place to do
research, behind only Toolik Field Station in Alaska. Nikita later
confided that he felt quite competitive with Toolik. Being far less
remote, the Alaskan station offers scientists considerable ameni-
ties, including seamless delivery from Amazon Prime. But Toolik
provides no alcohol, so Nikita balances its advantages by stocking
his station with Russian beer and crystal-blue bottles of Siberian
vodka, shipped into Cherskiy at a heavy cost. The drinks are often
consumed late at night in a roomy riverside sauna, under a sky
streaked pink by the midnight sun.

Nikita is the life of the station. He is at every meal, and any travel, by land or water, must be coordinated through him. His father is harder to find. One night, I caught Sergey alone in the dining room, having a late dinner. Squat and barrel-chested, he was sitting at a long table, his thick, gray rope of a ponytail hanging past his tailbone. His beard was a white Brillo Pad streaked with yellow. He chain-smoked all through the meal, drinking vodka, telling stories, and arguing about Russo-American relations. He kept insisting, loudly and in his limited English, that Donald Trump would be elected president in a few months. (Nikita would later tell me that Sergey has considered himself something of a prophet ever since he predicted the fall of the Soviet Union.) Late in the night, he finally mellowed when he turned to his favorite subjects, the deep past and far future of humankind. Since effectively handing the station over to his son, Sergey seems to have embraced a new role. He has become the station's resident philosopher.

Nikita would probably think *philosopher* too generous. "My dad likes to lie on the sofa and do science while I do all the work," he told me the next day. We were descending into an ice cave in Pleistocene Park. Step by cautious step, we made our way down a pair of rickety ladders that dropped 80 feet through the permafrost to the cave's bottom. Each time our boots found the next rung, we came eye to eye with a more ancient stratum of chilled soil. Even in the Arctic summer, temperatures in the underground network of chambers were below freezing, and the walls were coated with white ice crystals. I felt like we were wandering around in a geode.

Not every wall sparkled with fractals of white frost. Some were windows of clear ice, revealing mud that was 10,000, 20,000, even 30,000 years old. The ancient soil was rich with tiny bone fragments from horses, bison, and mammoths, large animals that would have needed a prolific, cold-resistant food source to survive the Ice Age Arctic. Nikita knelt and scratched at one of the frozen panels with his fingernail. Columns of exhaled steam floated up through the white beam of his headlamp. "See this?" he said. I leaned in, training my lamp on his thumb and forefinger. Between them, he held a thread of vegetable matter so tiny and pale that an errant breath might have reduced it to powder. It was a 30,000-year-old root that had once been attached to a bright-green blade of grass.

For the vast majority of the Earth's 4.5 billion spins around the sun, its exposed, rocky surfaces lay barren. Plants changed that.

Born in the seas like us, they knocked against the planet's shores for eons. They army-crawled onto the continents, anchored themselves down, and began testing new body plans, performing, in the process, a series of vast experiments on the Earth's surface. They pushed whole forests of woody stems into the sky to stretch their light-drinking leaves closer to the sun. They learned how to lure pollinators by unfurling perfumed blooms in every color of the rainbow. And nearly 70 million years ago, they began testing a new form that crept out from the shadowy edges of the forest and began spreading a green carpet of solar panel across the Earth.

For tens of millions of years, grasses waged a global land war against forests. According to some scientists, they succeeded by making themselves easy to eat. Unlike other plants, many grasses don't expend energy on poisons, or thorns, or other herbivore-deterring technologies. By allowing themselves to be eaten, they partner with their own grazers to enhance their ecosystem's nutrient flows.

Temperate-zone biomes can't match the lightning-fast biocycling of the tropics, where every leaf that falls to the steamy jungle floor is set upon by microbial swarms that dissolve its constituent parts. In a pine forest, a fallen branch might keep its nutrients locked behind bark and needle for years. But grasslands are able to keep nutrients moving relatively quickly, because grasses so easily find their way into the hot, wet stomachs of large herbivores, which are even more microbe-rich than the soil of the tropics. A grazing herbivore returns nutrients to the soil within a day or two, its thick, paste-like dung acting as a fertilizer to help the bitten blades of grass regrow from below. The blades sprout as if from everlasting ribbon dispensers, and they grow faster than any other plant group on Earth. Some bamboo grasses shoot out of the ground at a rate of several feet a day.

Grasses became the base layer for some of the Earth's richest ecosystems. They helped make giants out of the small, burrowing mammals that survived the asteroid that killed off the dinosaurs some 66 million years ago. And they did it in some of the world's driest regions, such as the sunbaked plains of the Serengeti, where more than 1 million wildebeests still roam. Or the northern reaches of Eurasia during the most-severe stretches of the Pleistocene.

The root between Nikita's thumb and forefinger was one foot

soldier among trillions that fought in an ecological revolution that human beings would come to join. We descended, after all, from tree-dwellers. Our nearest primate relatives, chimpanzees, bonobos, and gorillas, are still in the forest. Not human beings. We left Africa's woodlands and wandered into the alien ecology of its grassland savannas, as though sensing their raw fertility. Today, our diets—and those of the animals we domesticated—are still dominated by grasses, especially those we have engineered into mutant strains: rice, wheat, corn, and sugarcane.

"Ask any kid 'Where do animals live?' and they will tell you 'The forest,'" Nikita told me. "That's what people think of when they think about nature. They think of birds singing in a forest. They should think of the grassland."

Nikita and I climbed out of the ice cave and headed for the park's grassland. We had to cross a muddy drainage channel that he had bulldozed to empty a nearby lake, so that grass seeds from the park's existing fields could drift on the wind and fall onto the newly revealed soil. Fresh tufts of grass were already erupting out of the mud. Nikita does most of his violent gardening with a forest-mowing transporter on tank treads that stands more than 10 feet tall. He calls it the "mama mammoth."

When I first laid eyes on Pleistocene Park, I wondered whether it was the grassland views that first lured humans out of the woods. In the treeless plains, an upright biped can see almost into eternity. Cool Arctic winds rushed across the open landscape, fluttering its long ground layer of grasses. On the horizon, I made out a herd of large, gray-and-white animals. Their features came into focus as we hiked closer, especially after one broke into a run. They were horses, like those that sprinted across the plains of Eurasia and the Americas during the Pleistocene, their hooves hammering the ground, compressing the snow so that other grazers could reach cold mouthfuls of grass and survive the winter.

Like America's mustangs, Pleistocene Park's horses come from a line that was once domesticated. But it was hard to imagine these horses being tamed. They moved toward us with a boldness you don't often see in pens and barns. Nikita is not a man who flinches easily, but he backpedaled quickly when the horses feinted in our direction. He stooped and gathered a bouquet of grass and extended it tentatively. The horses snorted at the offer. They stared at us, dignified and curious, the mystery of animal consciousness beaming out from the black sheen of their eyes. At one point, four lined up in profile, like the famous quartet of gray horses painted by torchlight on the ceiling of Chauvet Cave, in France, some 30,000 years ago.

We walked west through the fields, to where a lone bison was grazing. When seen without a herd, a bison loses some of its glamour as a pure symbol of the wild. But even a single hungry specimen is an ecological force to be reckoned with. This one would eat through acres of grass by the time the year was out. In the warmer months, bison expend some of their awesome muscular energy on the destruction of trees. They shoulder into stout trunks, rubbing them raw and exposing them to the elements. It was easy to envision huge herds of these animals clearing the steppes of Eurasia and North America during the Pleistocene. This one had trampled several of the park's saplings, reducing them to broken, leafless nubs. Nikita and I worried that the bison would trample us, too, when, upon hearing us inch closer, he reared up his mighty, horned head, stilled his swishing tail, and stared, as though contemplating a charge.

We stayed low and headed away to higher ground to see a musk ox, a grazer whose entire being, inside and out, seems to have been carved by the Pleistocene. A musk ox's stomach contains exotic microbiota that are corrosive enough to process tundra scrub. Its dense layers of fur provide a buffer that allows it to graze in perfect comfort under the dark, aurora-filled sky of the Arctic winter, untroubled by skin-peeling, 70-below winds.

Nikita wants to bring hordes of musk oxen to Pleistocene Park. He acquired this one on a dicey boat ride hundreds of miles north into the ice-strewn Arctic Ocean. He would have brought back several others, too, but a pair of polar bears made off with them. Admiring the animal's shiny, multicolored coat, I asked Nikita whether he worried about poachers, especially with a depressed mining town nearby. He told me that hunters from Cherskiy routinely hunt moose, reindeer, and bear in the surrounding forests, "but they don't hunt animals in the park."

"Why?" I asked.

"Personal relationships," he said. "When the leader of the local mafia died, I gave the opening remarks at his funeral."

Filling Pleistocene Park with giant herbivores is a difficult task because there are so few left. When modern humans walked out of Africa, some 70,000 years ago, we shared this planet with more than 30 land-mammal species that weighed more than a ton. Of those animals, only elephants, hippos, rhinos, and giraffes remain. These African megafauna may have survived contact with human beings because they evolved alongside us over millions of years—long enough for natural selection to bake in the instincts required to share a habitat with the most dangerous predator nature has yet manufactured.

The giant animals that lived on other continents had no such luxury. When we first wandered into their midst, they may have misjudged us as small, harmless creatures. But by the time humans arrived in southern Europe, we'd figured out how to fan out across grasslands in small, fleet-footed groups. And we were carrying deadly projectiles that could be thrown from beyond the intimate range of an animal's claws or fangs.

Most ecosystems have checks against runaway predation. Population dynamics usually ensure that apex predators are rare. When Africa's grazing populations dip too low, for instance, lions go hungry and their numbers plummet. The same is true of sharks in the oceans. But when human beings' favorite prey thins out, we can easily switch to plant foods. This omnivorous resilience may explain a mystery that has vexed fossil hunters for more than a century, as they have slowly unearthed evidence of an extraordinary die-off of large animals all over the world, right at the end of the Pleistocene.

Some scientists think that extreme climate change was the culprit: the global melt transformed land-based biomes, and lumbering megafauna were slow to adapt. But this theory has weaknesses. Many of the vanished species had already survived millions of years of fluctuations between cold and warmth. And with a climate-caused extinction event, you'd expect the effects to be distributed across size and phylum. But small animals mostly survived the end of the Pleistocene. The species that died in high numbers were mammals with huge stores of meat in their flanks—precisely the sort you'd expect spear-wielding humans to hunt.

Climate change may have played a supporting role in these extinctions, but as our inventory of fossils has grown, it has strengthened the case for extermination by human rampage. Most telling is the timeline. Between 40,000 and 60,000 years ago, during an

ocean-lowering glaciation, a small group of humans set out on a
sea voyage from Southeast Asia. In only a few thousand years, they
skittered across Indonesia and the Philippines, until they reached
Papua New Guinea and Australia, where they found giant kangaroos, lizards twice as long as Komodo dragons, and furry, hippo-
sized wombats that kept their young in huge abdominal pouches.
Estimating extinction dates is tricky, but most of these species
seem to have vanished shortly thereafter.

It took at least another 20,000 years for human beings to trek
over the Bering land bridge to the Americas, and a few thousand
more to make it down to the southern tip. The journey seems to
have taken the form of an extended hunting spree. Before humans
arrived, the Americas were home to mammoths, bear-sized beavers, car-sized armadillos, giant camels, and a bison species twice
as large as those that graze the plains today. The smaller, surviving
bison is now the largest living land animal in the Americas, and it
barely escaped extermination: the invasion of gun-toting Europe-
ans reduced its numbers from more than 30 million to fewer than
2,000.

The pattern that pairs human arrival with megafaunal extinction is clearest in the far-flung islands that no human visited until
relatively recently. The large animals of Hawaii, Madagascar, and
New Zealand disappeared during the past 2,000 years, usually
within centuries of human arrival. This pattern even extends to
ocean ecosystems. As soon as industrial shipbuilding allowed large
groups of humans to establish a permanent presence on the seas,
we began hunting marine megafauna for meat and lamp oil. Less
than a century later, North Atlantic gray whales were gone, along
with 95 percent of North Atlantic humpbacks. Not since the asteroid struck have large animals found it so difficult to survive on
planet Earth.

In nature, no event happens in isolation. A landscape that loses
its giants becomes something else. Nikita and I walked all the way
to the edge of Pleistocene Park, to the border between the grassy
plains and the forest, where a line of upstart saplings was shooting
out of the ground. Trees like these had sprung out of the soils of
the northern hemisphere for ages, but until recently, many were
trampled or snapped in half by the mighty, tusked force of the
woolly mammoth.

It was only 3 million years ago that elephants left Africa and swept across southern Eurasia. By the time they crossed the land bridge to the Americas, they'd grown a coat of fur. Some of them would have waded into the shallow passes between islands, using their trunks as snorkels. In the deserts south of Alaska, they would have used those same trunks to make mental scent maps of water resources, which were probably sharper in resolution than a bloodhound's.

The mammoth family assumed new forms in new habitats, growing long fur in northern climes and shrinking to pygmies on Californian islands where food was scarce. But mammoths were always a keystone species on account of their prodigious grazing, their well-digging, and the singular joy they seemed to derive from knocking down trees. A version of this behavior is on display today in South Africa's Kruger National Park, one of the only places on Earth where elephants live in high densities. As the population has recovered, the park's woodlands have thinned, just as they did millions of years ago, when elephants helped engineer the African savannas that made humans into humans.

I have often wondered whether the human who first encountered a mammoth retained some cultural memory of its African cousin, in song or story. In the cave paintings that constitute our clearest glimpse into the prehistoric mind, mammoths loom large. In a single French cave, more than 150 are rendered in black outline, their tusks curving just so. In the midst of the transition from caves to constructed homes, some humans lived *inside* mammoths: 15,000 years ago, early architects built tents from the animals' bones and tusks.

Whatever wonderment human beings felt upon sighting their first mammoth, it was eventually superseded by more-practical concerns. After all, a single cold-preserved carcass could feed a tribe for a few weeks. It took less than 50 millennia for humans to help kill off the mammoths of Eurasia and North America. Most were dead by the end of the Ice Age. A few survived into historical times, on remote Arctic Ocean outposts like St. Paul Island, a lonely dot of land in the center of the Bering Sea where mammoths lived until about 3600 B.C. A final group of survivors slowly wasted away on Wrangel Island, just north of Pleistocene Park. Mammoth genomes tell us they were already inbreeding when the end came, around 2000 B.C. No one knows how the last mammoth

died, but we do know that humans made landfall on Wrangel Island around the same time.

The mammoth's extinction may have been our original ecological sin. When humans left Africa 70,000 years ago, the elephant
family occupied a range that stretched from that continent's
southern tip to within 600 miles of the North Pole. Now elephants
are holed up in a few final hiding places, such as Asia's dense forests. Even in Africa, our shared ancestral home, their populations
are shrinking, as poachers hunt them with helicopters, GPS, and
night-vision goggles. If you were an anthropologist specializing in
human ecological relationships, you might well conclude that one
of our distinguishing features as a species is an inability to coexist
peacefully with elephants.

But nature isn't fixed, least of all human nature. We may yet
learn to live alongside elephants, in all their spectacular variety.
We may even become a friend to these magnificent animals. Already, we honor them as a symbol of memory, wisdom, and dignity.
With luck, we will soon re-extend their range to the Arctic.

"Give me 100 mammoths and come back in a few years," Nikita
told me as he stood on the park's edge, staring hard into the fast-
growing forest. "You won't recognize this place."

The next morning, I met Sergey Zimov on the dock at the Northeast Science Station. In winter, when Siberia ices over, locals make
long-distance treks on the Kolyma's frozen surface, mostly in heavy
trucks, but also in the ancestral mode: sleighs pulled by fleet-
footed reindeer. (Many far-northern peoples have myths about flying reindeer.) Sergey and I set out by speedboat, snaking our way
down from the Arctic Ocean and into the Siberian wilderness.

Wearing desert fatigues and a black beret, Sergey smoked as he
drove, burning through a whole pack of unfiltered cigarettes. The
twin roars of wind and engine forced him to be even louder and
more aphoristic than usual. Every few miles, he would point at the
young forests on the shores of the river, lamenting their lack of
animals. "This is not wild!" he would shout.

It was early afternoon when we arrived at Duvanny Yar, a massive cliff that runs for six miles along the riverbank. It was like no
other cliff I'd ever seen. Rising 100 feet above the shore, it was a
concave checkerboard of soggy mud and smooth ice. Trees on its
summit were flopping over, their fun-house angles betraying the

thaw beneath. Its aura of apocalyptic decay was enhanced by the sulfurous smell seeping out of the melting cliffside. As a long seam of exposed permafrost, Duvanny Yar is a vivid window into the brutal geological reality of climate change.

Many of the world's far-northern landscapes, in Scandinavia, Canada, Alaska, and Siberia, are wilting like Duvanny Yar is. When Nikita and I had driven through Cherskiy, the local mining town, we'd seen whole houses sinking into mud formed by the big melt. On YouTube, you can watch a researcher stomp his foot on Siberian scrubland, making it ripple like a water bed. The northern reaches of the taiga are dimpled with craters hundreds of feet across, where frozen underground soil has gone slushy and collapsed, causing landslides that have sucked huge stretches of forest into the Earth. The local Yakutians describe one of the larger sinkholes as a "gateway to the underworld."

As the Duvanny Yar cliffside slowly melts into the Kolyma River, it is spilling Ice Age bones onto the riverbank, including woolly-rhino ribs and mammoth tusks worth thousands of dollars. A team of professional ivory hunters had recently picked the shore clean, but for a single 30-inch section of tusk spotted the previous day by a lucky German scientist. He had passed it around the dinner table at the station. Marveling at its smooth surface and surprising heft, I'd felt, for a moment, the instinctive charge of ivory lust, that peculiar human longing that has been so catastrophic for elephants, furry and otherwise. When I joked with Sergey that fresh tusks may soon be strewn across this riverbank, he told me he hoped he would be alive when mammoths return to the park.

The first of the resurrected mammoths will be the loneliest animal on Earth. Elephants are extremely social. When they are removed from normal herd life to a circus or a zoo, some slip into madness. Mothers even turn on their young.

Elephants are matriarchal: males generally leave the herd in their teens, when they start showing signs of sexual maturity. An elephant's social life begins at birth, when a newborn calf enters the world to the sound of joyous stomping and trumpeting from its sisters, cousins, aunts, and, in some cases, a grandmother.

Mammoth herds were likewise matriarchal, meaning a calf would have received patient instruction from its female elders. It would have learned how to use small sticks to clean dirt from the cracks in its feet, which were so sensitive that they could feel the steps of a distant herd member. It would have learned how to wield a trunk stuffed with more muscles than there are in the entire human body, including those that controlled its built-in water hose. It would have learned how to blast trumpet notes across the plains, striking fear into cave lions, and how to communicate with its fellow herd members in a rich range of rumbling sounds, many inaudible to the human ear.

The older mammoths would have taught the calf how to find ancestral migration paths, how to avoid sinkholes, where to find water. When a herd member died, the youngest mammoth would have watched the others stand vigil, tenderly touching the body of the departed with their trunks before covering it with branches and leaves. No one knows how to recreate this rich mammoth culture, much less how to transmit it to that cosmically bewildered first mammoth.

Or to an entire generation of such mammoths. The Zimovs won't be able to slow the thawing of the permafrost if they have to wait for their furry elephant army to grow organically. That would take too long, given the species' slow breeding pace. George Church, the Harvard geneticist, told me he thinks the mammoth-manufacturing process can be industrialized, complete with synthetic-milk production, to create a seed population that numbers in the tens of thousands. But he didn't say who would pay for it—at the Northeast Science Station, there was open talk of recruiting a science-friendly Silicon Valley billionaire—or how the Zimovs would deploy such a large group of complex social animals that would all be roughly the same age.

Nikita and Sergey seemed entirely unbothered by ethical considerations regarding mammoth cloning or geoengineering. They saw no contradiction between their veneration of "the wild" and their willingness to intervene, radically, in nature. At times they sounded like villains from a Michael Crichton novel. Nikita suggested that such concerns reeked of a particularly American piety. "I grew up in an atheist country," he said. "Playing God doesn't bother me in the least. We are already doing it. Why not do it better?"

Sergey noted that other people want to stop climate change by putting chemicals in the atmosphere or in the ocean, where they could spread in dangerous ways. "All I want to do is bring animals back to the Arctic," he said.

As Sergey and I walked down the riverbank, I kept hearing a cracking sound coming from the cliff. Only after we stopped did I register its source, when I looked up just in time to see a small sheet of ice dislodge from the cliffside. Duvanny Yar was bleeding into the river before our very eyes.

In 1999, Sergey submitted a paper to the journal *Science* arguing that Beringian permafrost contained rich "yedoma" soils left over from Pleistocene grasslands. (In other parts of the Arctic, such as Norway and eastern Canada, there is less carbon in the permafrost; if it thaws, sea levels will rise, but much less greenhouse gas will be released into the atmosphere.) When Beringia's pungent soils are released from their icy prison, microbes devour the organic contents, creating puffs of carbon dioxide. When this process occurs at the bottom of a lake filled with permafrost melt, it creates bubbles of methane that float up to the surface and pop, releasing a gas whose greenhouse effects are an order of magnitude worse than carbon dioxide's. Already more than 1 million of these lakes dot the Arctic, and every year, new ones appear in NASA satellite images, their glimmering surfaces steaming methane into the closed system that is the Earth's atmosphere. If huge herds of megafauna recolonize the Arctic, they too will expel methane, but less than the thawing frost, according to the Zimovs' estimates.

Science initially rejected Sergey's paper about the danger posed by Beringia's warming. But in 2006, an editor from the journal asked Sergey to resubmit his work. It was published in June of that year. Thanks in part to him, we now know that there is more carbon locked in the Arctic permafrost than there is in all the planet's forests and the rest of the atmosphere combined.

For my last day in the Arctic, Nikita had planned a send-off. We were to make a day trip, by car, to Mount Rodinka, on Cherskiy's outskirts. Sergey came along, as did Nikita's daughters and one of the German scientists.

Rodinka is referred to locally as a mountain, though it hardly merits the term. Eons of water and wind have rounded it down to a dark, stubby hill. But in Siberia's flatlands, every hill is a mountain. Halfway up to the summit, we already had a God's-eye view of the surrounding landscape. The sky was lucid blue but for a thin mist that hovered above the Kolyma River, which slithered, through a mix of evergreens and scrub, all the way to the horizon. At the foot of the mountain, the gold-mining town and its airstrip hugged the river. In the dreamy, deep-time atmosphere of Pleistocene Park, it had been easy to forget this modern human world outside the park's borders.

Just before the close of the nineteenth century, in the pages of this magazine, John Muir praised the expansion of Yellowstone, America's first national park. He wrote of the forests, yes, but also of the grasslands, the "glacier meadows" whose "smooth, silky lawns" pastured "the big Rocky Mountain game animals." Already the park had served "the furred and feathered tribes," he wrote. Many were "in danger of extinction a short time ago," but they "are now increasing in numbers."

Yellowstone's borders have since been expanded even farther. The park is now part of a larger stretch of land cut out from ranches, national forests, wildlife refuges, and even tribal lands. This Greater Yellowstone Ecosystem is 10 times the size of the original park, and it's home to the country's most populous wild-bison herd. There is even talk of extending a wildlife corridor to the north, to provide animals safe passage between a series of wilderness reserves, from Glacier National Park to the Canadian Yukon. But not everyone supports Yellowstone's outward expansion. The park is also home to a growing population of grizzly bears, and they have started showing up in surrounding towns. Wolves were reintroduced in 1995, and they, too, are now thriving. A few have picked off local livestock.

Sergey sees Pleistocene Park as the natural next step beyond Yellowstone in the rewilding of the planet. But if Yellowstone is already meeting resistance as it expands into the larger human world, how will Pleistocene Park fare if it leaves the Kolyma River basin and spreads across Beringia?

The park will need to be stocked with dangerous predators. When they are absent, herbivore herds spread out, or they feel safe enough to stay in the same field, munching away mindlessly until it's overgrazed. Big cats and wolves force groups of grazers into dense, watchful formations that move fast across a landscape, visiting a new patch of vegetation each day in order to mow it with their teeth, fertilize it with their dung, and trample it with their many-hooved plow. Nikita wants to bring in gray wolves, Siberian tigers, or cold-adapted Canadian cougars. If it becomes a trivial challenge to resurrect extinct species, perhaps he could even repopulate Siberia with cave lions and dire wolves. But what will happen when one of these predators wanders onto a city street for the first time?

"This is a part of the world where there is very little agriculture, and very few humans," Sergey told me. He is right that Beringia is sparsely populated, and that continuing urbanization will likely clear still more space by luring rural populations into the cities. But the region, which stretches across Alaska and the Canadian Yukon, won't be empty any time soon. Fifty years from now, there will still be mafia leaders to appease, not to mention indigenous groups and the governments of three nations, including two that spent much of the last century vying for world domination. America and Russia often cooperate in the interest of science, especially in extreme environments like Antarctica and low-Earth orbit, but the Zimovs will need a peace that persists for generations.

Sergey envisions a series of founding parks, "maybe as many as 10," scattered across Beringia. One would be along the Yentna River, in Alaska, another in the Yukon. A few would be placed to the west of Pleistocene Park, near the Ural mountain range, which separates Siberia from the rest of Russia. As Sergey spoke, he pointed toward each of these places, as if they were just over the horizon and not thousands of miles away.

Sergey's plan relies on the very climate change he ultimately hopes to forestall. "The top layer of permafrost will melt first," he said. "Modern ecosystems will be destroyed entirely. The trees will fall down and wash away, and grasses will begin to appear." The Mammoth Steppe would spread from its starting nodes in each park until they all bled into one another, forming a megapark that spanned the entire region. Humans could visit on bullet trains built on elevated tracks, to avoid disturbing the animals' free movement. Hunting could be allowed in designated areas. Gentler souls could go on Arctic safari tours.

When Sergey was out of earshot, I asked Nikita whether one of his daughters would one day take over Pleistocene Park to see this plan through. We were watching two of them play in an old Soviet-military radar station, about 100 yards from Rodinka's peak.

"I took the girls to the park last week, and I don't think they were too impressed," Nikita told me, laughing. "They thought the horses were unfriendly." I told him that wasn't an answer. "I'm not as selfish as my father," he said. "I won't force them to do this."

Before I left to catch a plane back to civilization, I stood with Sergey on the mountaintop once more, taking in the view. He had slipped into one of his reveries about grasslands full of animals. He seemed to be suffering from a form of solastalgia, a condition described by the philosopher Glenn Albrecht as a kind of existential grief for a vanished landscape, be it a swallowed coast, a field turned to desert, or a bygone geological epoch. He kept returning to the idea that the wild planet had been interrupted midway through its grand experiment, its 4.5-billion-year blending of rock, water, and sunlight. He seems to think that the Earth peaked during the Ice Age, with the grassland ecologies that spawned human beings. He wants to restore the biosphere to that creative summit, so it can run its cosmic experiment forward in time. He wants to know what new wonders will emerge. "Maybe there will be more than one animal with a mind," he told me.

I don't know whether Nikita can make his father's mad vision a material reality. The known challenges are immense, and there are likely many more that he cannot foresee. But in this brave new age when it is humans who make and remake the world, it is a comfort to know that people are trying to summon whole landscapes, Lazarus-like, from the tomb. "Come forth," they are saying to woolly mammoths. Come into this habitat that has been prepared for you. Join the wolves and the reindeer and the bison who survived you. Slip into your old Ice Age ecology. Wander free in this wild stretch of the Earth. Your kind will grow stronger as the centuries pass. This place will overflow with life once again. Our original sin will be wiped clean. And if, in doing all this, we can save our planet and ourselves, that will be the stuff of a new mythology.

JACQUELINE DETWILER

It'll Take an Army to Kill the Emperor

FROM *Popular Mechanics*

I. Precision Medicine — Or, What Is Cancer, Though, Really?

WHEN YOU VISIT St. Jude Children's Research Hospital in Memphis, Tennessee, you expect to feel devastated. It starts in the waiting room. *Oh, here we go with the little red wagons,* you think, observing the cattle herd of them rounded up by the entrance to the Patient Care Center. *Oh, here we go with the crayon drawings of needles.* The itch begins at the back of your throat, and you start blinking very fast and mentally researching how much money you could donate without starving. Near a row of arcade games, a pre-teen curls his face into his mother's shoulder while she strokes his head. *Oh, here we go.*

But the more time you spend at St. Jude, the more that feeling is replaced with wonder. In a cruel world you've found a free hospital for children, started by a Hollywood entertainer as a shrine to the patron saint of lost causes. There is no other place like this. Corporations that have nothing to do with cancer—nothing to do with medicine, even—have donated vast sums of money just to be a part of it. There's a Chili's Care Center. The cafeteria is named for Kay Jewelers.

Scott Newman's office is in the Brooks Brothers Computational Biology Center, where a team of researchers is applying computer science and mathematics to the question of why cancer happens to children. Like many computer people, Newman is very smart and a little quiet and doesn't always *exactly* meet your eyes when he speaks to you. He works on St. Jude's Genomes for Kids project, which invites newly diagnosed patients to have both their healthy and tumor cells genetically sequenced so researchers can poke around.

"Have you seen a circle plot before?" Newman asks, pulling out a diagram of the genes in a child's cancer. "If I got a tattoo, it would be one of these." Around the outside of the circle plot is something that looks like a colorful bar code. Inside, a series of city skylines. Through the center are colored arcs like those nail-and-string art projects students make in high school geometry class. The diagram represents everything that has gone wrong within a child's cells to cause cancer. It's beautiful.

"These are the genes in this particular tumor that have been hit," Newman says in a Yorkshire accent that emphasizes the *t* at the end of the word *hit* in a quietly violent way. "And that's just one type of thing that's going on. Chromosomes get gained or lost in cancer. This one has gained that one, that one, that one, that one," he taps the page over and over. "And then there are structural rearrangements where little bits of genome get switched around." He points to the arcs sweeping across the page. "There are no clearly defined rules."

It's not like you don't have cancer and then one day you just do. Cancer—or, really, cancers, because cancer is not a single disease—happens when glitches in genes cause cells to grow out of control until they overtake the body, like a kudzu plant. Genes develop glitches all the time: there are roughly 20,000 genes in the human body, any of which can get misspelled or chopped up. Bits can be inserted or deleted. Whole copies of genes can appear and disappear, or combine to form mutants. The circle plot Newman has shown me is not even the worst the body can do. He whips out another one, a snarl of lines and blocks and colors. This one would not make a good tattoo.

"As a tumor becomes cancerous and grows, it can accumulate many thousands of genetic mutations. When we do whole genome sequencing, we see all of them," Newman says. To whittle down the complexity, he applies algorithms that pop out gene mutations most likely to be cancer-related, based on a database of all the mutations researchers have already found. Then, a genome analyst manually determines whether each specific change the algorithm found seems likely to cause problems. Finally, the department brings its list of potentially important changes to a committee of St. Jude's top scientists to discuss and assign a triage score. The mutations that seem most likely to be important get investigated first.

It took 13 years and cost $2.7 billion to sequence the first genome, which was completed in 2003. Today, it costs $1,000 and takes less than a week. Over the last two decades, as researchers like Newman have uncovered more and more of the individual genetic malfunctions that cause cancer, teams of researchers have begun to tinker with those mutations, trying to reverse the chaos they cause. (The first big success in precision medicine was Gleevec, a drug that treats leukemias that are positive for a common structural rearrangement called the Philadelphia chromosome. Its launch in 2001 was revolutionary.) Today, there are 11 genes that can be targeted with hyperspecific cancer therapies, and at least 30 more being studied. At Memorial Sloan Kettering Cancer Center in New York City, 30 to 40 percent of incoming patients now qualify for precision medicine studies.

Charles Mullighan, a tall, serious Australian who also works at St. Jude, is perhaps the ideal person to illustrate how difficult it will be to cure cancer using precision medicine. After patients' cancer cells are sequenced, and the wonky mutations identified, Mullighan's lab replicates those mutations in mice, then calls St. Jude's chemical library to track down molecules—some of them approved medicines from all over the world, others compounds that can illuminate the biology of tumors—to see if any might help.

If Mullighan is lucky, one of the compounds he finds will benefit the mice, and he'll have the opportunity to test it in humans. Then he'll hope there are no unexpected side effects, and that the cancer won't develop resistance, which it often does when you futz with genetics. There are about 20 subtypes of the leukemia Mullighan studies, and that leukemia is one of a hundred different subtypes of cancer. This is the kind of precision required in precision cancer treatment—even if Mullighan succeeds in identifying a treatment that works as well as Gleevec, with the help of an entire, well-funded hospital, it still will work for only a tiny proportion of patients.

Cancer is not an ordinary disease. Cancer is *the* disease—a phenomenon that contains the whole of genetics and biology and human life in a single cell. It will take an army of researchers to defeat it.

Luckily, we've got one.

Interlude

"I used to do this job out in L.A.," says the attendant at the Hertz counter at Houston's George Bush Intercontinental Airport. "There, everyone is going on vacation. They're going to the beach or Disneyland or Hollywood or wherever.

"Because of MD Anderson, I see more cancer patients here. They're so skinny. When they come through this counter, they're leaning on someone's arm. They can't drive themselves. You think, there is no way this person will survive. And then they're back in three weeks, and in six months, and a year. I'm sure I miss some, who don't come through anymore because they've died. But the rest? They come back."

II. Checkpoint Inhibitor Therapy—Or, You Have the Power Within You!

On a bookshelf in Jim Allison's office at MD Anderson Cancer Center in Houston (and on the floor surrounding it) are so many awards that some still sit in the boxes they came in. The Lasker-DeBakey Clinical Medical Research Award looks like the *Winged Victory* statue in the Louvre. The Breakthrough Prize in Life Sciences, whose benefactors include Sergey Brin, Anne Wojcicki, and Mark Zuckerberg, came with $3 million.

"I gotta tidy that up sometime," Allison says.

Allison has just returned to the office from back surgery that fused his L3, L4, and L5 vertebrae, which has slightly diminished his Texas rambunctiousness. Even on painkillers, though, he can explain the work that many of his contemporaries believe will earn him the Nobel Prize: he figured out how to turn the immune system against tumors.

Allison is a basic scientist. He has a Ph.D., rather than an M.D., and works primarily with cells and molecules rather than patients. When T-cells, the most powerful "killer cells" in the immune system, became better understood in the late 1960s, Allison became fascinated with them. He wanted to know how it was possible that a cell roaming around your body knew to kill infected cells but not healthy ones. In the mid-1990s, both Allison's lab and the lab of Jeffrey Bluestone at the University of Chicago noticed that a molecule called CTLA-4 acted as a brake on T-cells, preventing them from wildly attacking the body's own cells, as they do in autoimmune diseases.

Allison's mother died of lymphoma when he was a child, and he has since lost two uncles and a brother to the disease. "Every time I found something new about how the immune system works, I would think, *I wonder how this works on cancer?*" he says. When the scientific world discovered that CTLA-4 was a brake, Allison alone wondered if it might be important in cancer treatment. He launched an experiment to see if blocking CTLA-4 would allow the immune system to attack cancer tumors in mice. Not only did the mice's tumors disappear, the mice were thereafter immune to cancer of the same type.

Ipilimumab ("ipi" for short) was the name a small drug company called Medarex gave the compound it created to shut off CTLA-4 in humans. Early trials of the drug, designed just to show whether ipi was safe, succeeded so wildly that Bristol-Myers Squibb bought Medarex for $2.4 billion. Ipilimumab (now marketed as Yervoy) became the first "checkpoint inhibitor": it blocks one of the brakes, or checkpoints, the immune system has in place to prevent it from attacking healthy cells. Without the brakes the immune system can suddenly, incredibly, recognize cancer as the enemy.

"You see the picture of that woman over there?" Allison points over at his desk. Past his lumbar-support chair, the desk is covered in papers and awards and knickknacks and frames, including one containing a black card with the words "Never never never give up" printed on it. Finally, the photo reveals itself, on a little piece of blue card stock.

"That's the first patient I met," Allison says. "She was about twenty-four years old. She had metastatic melanoma. It was in her brain, her lungs, her liver. She had failed everything. She had just graduated from college, just gotten married. They gave her a month."

The woman, Sharon Belvin, enrolled in a phase-two trial of ipi-

limumab at Memorial Sloan Kettering, where Allison worked at
the time. Today, Belvin is thirty-five, cancer-free, and the mother
of two children. When Allison won the Lasker prize, in 2015, the
committee flew Belvin to New York City with her husband and
her parents to see him receive it. "She picked me up and started
squeezing me," Allison says. "I walked back to my lab and thought,
Wow, I cure mice of tumors and all they do is bite me." He adds, dryly,
"Of course, we gave them the tumors in the first place."

After ipi, Allison could have taken a break and waited for his
Nobel, driving his Porsche Boxster with the license plate CTLA-4
around Houston and playing the occasional harmonica gig. (Allison, who grew up in rural Texas, has played since he was a teenager and once performed "Blue Eyes Crying in the Rain" onstage
with Willie Nelson.) Instead, his focus has become one of two serious problems with immunotherapy: it only works for some people.

So far, the beneficiaries of immune checkpoint therapy appear
to be those with cancer that develops after repeated genetic mutations—metastatic melanoma, non-small-cell lung cancer, and bladder cancer, for example. These are cancers that often result from
bad habits like smoking and sun exposure. But even within these
types of cancer, immune checkpoint therapies improve long-term
survival in only about 20 to 25 percent of patients. In the rest, the
treatment fails, and researchers have no idea why.

Lately, Allison considers immune checkpoint therapy a "platform"—a menu of treatments that can be amended and combined
to increase the percentage of people for whom it works. A newer
drug called Keytruda that acts on a different immune checkpoint,
PD-1, knocked former president Jimmy Carter's metastatic melanoma into remission in 2015. Recent trials that blocked both PD-1
and CTLA-4 in combination improved long-term survival in 60
percent of melanoma patients. Now, doctors are combining checkpoint therapies with precision cancer drugs, or with radiation, or
with chemotherapy. Allison refers to this as "one from column A,
and one from column B."

The thing about checkpoint inhibitor therapy that is so exciting
—despite the circumscribed group of patients for whom it works,
and despite sometimes mortal side effects from the immune system going buck-wild once the brakes come off—is the length of
time it can potentially give people. Before therapies that exploited
the immune system, response rates were measured in a few extra

months of life. Checkpoint inhibitor therapy helps extremely sick people live for *years*. So what if it doesn't work for everyone? Every cancer patient you can add to the success pile is essentially cured.

Jennifer Wargo is another researcher at MD Anderson who is trying to predict who will respond to checkpoint inhibitor therapy and who will not. Originally a nurse, Wargo got so interested in biology that she went back to school for a bachelor's degree, then a medical degree, and then a surgical residency at Harvard. It was during her first faculty position, also at Harvard, around 2008, that she started to wonder how the microbiome—the bacteria that live in the human body, of which there are roughly 40 trillion in the average 155-pound man—might affect cancer treatment. Wargo was investigating the bacteria that live near the site of pancreatic cancer—in and around the tumor. Could you target those bacteria with drugs and make the cancer recede more quickly?

In the early 2010s, research about the microbiome in the human gut—the bacteria in humans' stomachs and intestines that appear to affect immune function, gene expression, and mood, among other things—gained traction in journals. Before long, two separate researchers had shown that you could change a mouse's response to immune checkpoint inhibitor therapy by giving him certain kinds of bacteria. Wargo added the microbiome to her slate of experiments. Along with her team, she collected gut microbiome samples from more than 300 cancer patients who then went on to receive checkpoint inhibitors as treatment. The results were, Wargo says, "night and day." People who had a higher diversity of gut bacteria had a stronger response to checkpoint inhibitor therapy.

Now, Wargo is transplanting stool samples from patients into germ-free mice with melanoma, to see if she can predict whether the mice will mimic the treatment responses of the people whose bacteria they received. "Can we change the gut microbiome to enhance responses to therapy . . . or even prevent cancer altogether?" she says. "Ah god, that would be the holy grail, wouldn't it?" she whispers, as if not to invite bad luck. "It's gonna take a lot of work to get there, but I think the answer is gonna be yes."

Immunotherapies do have one other problem worth worrying about, one that underlies the most frustrating experience of having cancer. When a patient is diagnosed, the first therapy is still one of the standards: surgery, radiation, or chemotherapy. Cut, burn,

or poison, as the doctors say. Doctors can't use promising immunotherapies as first-line treatments yet because immunotherapies
are still dangerous: No one knows what will happen long-term if
you shut off the immune system's brakes. Does a patient survive
cancer just to develop another terrible disease, like amyotrophic
lateral sclerosis (ALS), in 15 years?

Interlude

*"Just to play devil's advocate," says a woman at a margarita bar and
restaurant in Santa Fe, New Mexico. "Don't you think the cure exists somewhere already and the medical industrial complex is hiding it? People stand
to lose billions of dollars. Don't you think they want to keep that money?"*

*I have been talking to this woman for 20 minutes. She is familiar with
cancer. She works with natural cures, is a big fan of neuroscience, and
knows some of the prominent names in medical research. I tell her that the
conspiracy theory she is referencing—that the government or pharmaceutical industry is hiding the cure for cancer—can't be true. Of course it's hard
to believe that Richard Nixon initiated the war on cancer in 1971 and the
disease still kills 595,690 people a year. And that the most brilliant minds
of our time have turned HIV into a chronic disease but cancer continues
relatively unchecked. And yet I've talked to 35 researchers and policymak-
ers and visited seven cancer centers and I haven't seen a shred of evidence
that doctors who treat very sick people—and whose job it is, sometimes, to
tell people that they will die—aren't trying with their very souls to succeed
at their jobs.*

"It's just that it's hard," I say.

*The woman huffs. Someone more interesting is sitting on the other side
of her. And that's the end of that.*

III. CAR-T Cells—Or, Tiny Machines

On a shelf in Crystal Mackall's office at Stanford University in Palo
Alto, California, catty-corner to a window that looks out on a lovely
California scrub scape, is a teddy bear that once belonged to a boy
named Sam.* Sam, who Mackall treated at the National Cancer In-

---

* Some names have been changed to protect patient privacy.

stitute more than 10 years ago, had Ewing's sarcoma, a rare cancer that usually affects children and grows in or around bones.

Mackall is a pediatric oncologist with a dark blond bob and a wry, take-no-prisoners sense of humor. She has worked on cancer since the 1980s, so she has met a lot of very, very sick children. The way Mackall tells the story of Sam, like she's taking a shot of foul-tasting medicine, you can see the distance she's had to put between her emotions and her work. "We lost Sam. He was ten," she says. "We gave him immunotherapy and it didn't work."

With that, Mackall moves on to the story of a girl named Lisa, who is pictured in a photo not far from the bear. Lisa had the same illness as Sam around the same time, but her therapy *did* work. Lisa's story lasts more than a minute, with Mackall practically cheering at the end. "So she remained fertile and that's her little boy!" she yells, gesturing toward Lisa's photo. Mackall smiles the pained, confused smile of someone who has inexplicably survived a car crash. "You have your ups and your downs," she says.

Overall, children's cancer has been one of the great success stories in cancer treatment. In the 1970s, dramatic advances in chemotherapy put most patients with certain types of leukemia (particularly acute lymphoblastic leukemia in B cells, otherwise known as B-ALL) into remission. Today, 84 percent of children who get ALL can be cured. But then treatment stalled. "We have made steady progress, by all accounts," says Mackall. "But it's been largely incremental. And there've been these plateaus that have just driven us crazy."

In those unfortunate few children who relapsed or didn't respond to the chemo, or who got a different variety of cancer, like Ewing's sarcoma, there were few treatments left to try. Mackall's patients came to her after having had surgery and then chemotherapy once, twice, three times. "You can just see, they're beat-up. They're making it, but all they do is get their treatments," she says. "They didn't have enough energy to do anything else." And then, if they lasted long enough, they got into a trial.

There are several ways to turn the immune system against cancer. Checkpoint inhibitor therapy is one of them. But it doesn't work in all patients, especially children, whose cancers generally do not have the vast numbers of mutations needed to attract the attention of a newly brake-free immune system. For a long, dark time, immunotherapists would try other sorts of techniques to get the immune system to respond in these patients, and the patients would die anyway, like Sam did. The treatments were toxic or they damaged the brain or they just didn't work. The doctors would recommend hospice. Hospice. Hospice.

And then all the research began to pay off. In August 2010, a retired correctional officer named Bill Ludwig walked into the Hospital of the University of Pennsylvania to try a new therapy developed by a researcher named Carl June. Ludwig had chronic lymphocytic leukemia (CLL), another cancer that affects B-cells. Multiple rounds of chemotherapy had failed to cure it, and he didn't qualify for a bone marrow transplant. June's idea, which was so risky that the National Institutes of Health had turned down several grant applications to fund it, was the only option Ludwig had. June had only enough money to try it in three patients. Ludwig went first.

To understand how June's therapy works, consider the T-cells that Jim Allison found fascinating. They're cells that kill other cells, but they don't kill you because they have a built-in targeting mechanism. Each person has millions of T-cells, and each one of those T-cells matches a single virus, like a lock and a key. If a virus enters the body, its own personal T-cell key will find and destroy it, then copy and copy and copy itself until the virus succumbs. "I liken it to a bloodhound," says Mackall. "What the marker says to the T-cell is: Anything that has this thing on it, kill it."

Previously, researchers had created a fake key called a chimeric antigen receptor, or CAR, that matched a particular lock, CD19, on B-cells, which is where Ludwig's leukemia was. During the trial, Ludwig's doctors removed as many of Ludwig's T-cells as they could, and June's team inserted the CAR using a modified form of HIV, which can edit genes. Then they returned the T-cells to Ludwig.

Ten days later, Ludwig started to have chills and fever, like he had the flu. He was so ill that doctors moved him to the intensive care unit. But then, less than a month later, he was in remission. The T-cells had located and demolished the cancer, the same way they would a virus.

When case studies of the first three patients were published in scientific journals, mainstream media went crazy: "Cancer treated with HIV!" they shouted. But it was a later study that showed that the furor was warranted: when the Penn team partnered with the Children's Hospital of Philadelphia to try CAR-T cell therapy against B-ALL in children, the cancer disappeared in 24 out of 27 patients.

Novartis was the drug company that partnered with the University of Pennsylvania to turn June's treatment into a drug for the general public, and the company submitted results of all three required levels of tests to the Food and Drug Administration early this year. If the FDA approves the drug, any child who has B-ALL and has failed her first therapy can have her white blood cells removed, frozen, and shipped to Novartis's processing facility in Morris Plains, New Jersey, where molecular engineers will insert the new "key" and send the T-cells back. The patient gets a one-time infusion, and there's an 83 percent chance she will be cured.

"We also do a second measure of remissions where we look to see if there's any measurable disease at all," says David Lebwohl, Novartis's global program head for CAR-T treatments. "A more sensitive test than just looking in the blood. And that was also negative for 83 percent of the patients."

An 83 percent cure rate in children who would otherwise die is a monumental achievement. If there is a moment where a culture hits on an idea that can cure a disease—vaccines, for example, or penicillin—we are in it. It is difficult to overstate this: humans have been trying to create a cell therapy for cancer patients for generations. "People said: That can't be done, You can't make them from cancer patients, You can't make them, You can't get them, It's too complicated," says Crystal Mackall. "But it's happening." Though Novartis couldn't confirm an official release date, Mackall suspects the drug will become widely available this year.

Cancer being cancer, of course, there are limitations: until it clears further FDA hurdles, Novartis's drug will be available only for children with B-ALL and not for any of the dozens of other types of cancers that affect children and adults. In solid tumors, the CAR-T cells aren't strong enough to kill the whole thing, or they die before they finish the job. Worse, once attacked, some leukemia cells will remove their CD19 proteins and go back into hiding. "The thing about cancer is, it's quite a foe," Mackall says. "The minute you think you've got the one thing for it, it'll outsmart you."

Slowly, though, the successes are mounting. At City of Hope National Medical Center just outside Los Angeles, Behnam Badie, an

Iranian-born brain surgeon who has the kind of bedside manner you'd dream of if you ever required a brain surgeon, is developing a surgical device that can continuously infuse CAR-T cells into the brain tumors of cancer patients while he operates. For a while, he was working with the California Institute of Technology to build a magnetic helmet that could move the cells to the correct places, but the project ran out of money.

Meanwhile, Crystal Mackall is working on a backup target for the CAR-T cells, CD22, in case a child's cancer resists the ones targeted to CD19. She is also trying to make the cells live longer. Working with similar but slightly different engineered cells, she has managed to get her therapy to stay alive and working for up to two years in patients with solid sarcomas. One of her patients has since gotten married and bought a farm. Another went on a volunteer trip to Africa.

Mackall likens genetically engineered cells to rudimentary machines. Over the next decade, she says, scientists will refine them until they can control where they go and what they do and when. "We're going to be in a situation," she says, "where a doctor can tell a patient to take pills to activate his cells one week and then rest them the next." In fact, a biotech company based in San Diego called BioAtla has already developed conditionally active markers that could tell a T-cell to kill or not kill based on where it is in the body.

Eventually, programmable cell machines could fight autoimmune diseases, or arthritis. They could be used to rebuild collagen in athletes' knees. But, because such powerful new technology requires a ton of risk to attempt, none of this would have been developed without an adversary as vile as cancer to require it. "We treated 49 kids at the National Cancer Institute with refractory leukemia. Every single one of those kids had exhausted every other therapy available. If it weren't for the CAR-T cells, they were gonna die," Mackall says. Sixty percent of those children went into remission, and a sizable fraction of those appear to be cured. "You're able to take the chance only in that situation, when people don't have other options."

People will die waiting for CAR-T therapy to really, truly happen. In the United States, doctors aren't permitted to experiment on patients who have other options, and it will take a long time for CAR-T to prove itself better than the treatments already available.

But someone has to choose to take the first walk down the path to the future. In a final act that is equal parts self-preservation and sacrifice, that person is usually a cancer patient. And soon, more of them will be able to make the decision for themselves.

Interlude

"What're ya down here for?" asks an older gentleman at the bar of a tourist barbecue joint near my hotel in Memphis. I'm halfway through a plate of pickles and dry-rubbed ribs. I explain that I've spent all day at St. Jude.

"God bless you," he says. "I couldn't do it." The man is from Texas — he works in shipping or packing or something or other.

The bartender, a bubbly twenty-three-year-old, offers the gentleman another beer. "You know, I was treated at St. Jude. Diagnosed at ten. Cured at thirteen," he says, beaming.

"Was it awful?" I ask. "Getting cancer as a kid?"

"Naw, I loved going to St. Jude. I remember I looked forward to school being over so I could go over to the hospital and get chemo. Your doctors are so happy to see you."

The bartender is studying to be a truck driver so he can visit California. He's not sure if he'll settle down there, but it seems nice.

The man from Texas looks at the bartender hard for a good minute, says, "You're a lucky man, son."

IV. Postmodern Radiation — Or, Any Other Ideas?

To get to the Los Alamos National Laboratory in New Mexico, you drive from Santa Fe through peach-parfait mesas and off into the sunset. Even on the public roads, there are checkpoints where security officers will ask to see your driver's license. The deeper you go, the more intense the screening gets, until finally you end up in a place employees just call "behind the fence."

After the public roads but before "behind the fence" are the hot cells: four-foot by three-and-a-half-foot boxes where employees use robot hands controlled by joysticks to process non-weapons-grade isotopes. The isotopes are made on another mesa, by a linear particle accelerator that shoots rare metals with proton beams.

Just outside the hot cells, Eva Birnbaum, the isotope production facility's program manager, asks me if I know what a decay chain is. She points in the direction of an expanded periodic table that, despite a year of college chemistry, means about as much to me as a list of shipbuilding supplies from the 1600s. Birnbaum launches into a primer on radiochemistry: Isotopes are chemical elements with too many or too few neutrons in their centers. Some of these are unstable and therefore release energy by shooting out various types of particles. Unstable isotopes are radioactive, and the particles they shoot out are known as ionizing radiation.

As for what a decay chain is: when radioactive isotopes release radiation, they usually turn into another radioactive isotope, which releases radiation until it turns into another radioactive isotope, and so on, until it hits on something stable. The pattern by which a particular isotope morphs is its decay chain. Today, in addition to whatever goes on behind the fence, Los Alamos National Laboratory is the primary producer of certain isotopes whose decay chains make them useful for medical scans, such as PET scans and heart-imaging techniques. Scientists at Los Alamos deliver the parent isotope in a container called a cow. As the parent decays, doctors "milk" the daughter isotope off to image patients' hearts.

Decay chains present both an opportunity and a responsibility for the U.S. government. You can't just throw decaying radioactive isotopes into a landfill, so after the nuclear age and a half-century Cold War with the USSR, there are caches of radioactive uranium and plutonium isotopes sitting around gradually turning into other stuff. One of these caches is uranium-233, which was originally created for a reactor program and is currently stored at the Oak Ridge National Laboratory in Tennessee. Over the last 40-some years, it has been slowly turning into thorium-229.

Thorium-229's decay chain leads to actinium-225, which is of interest to cancer researchers for several reasons. For one thing, actinium-225's decay chain goes on for several generations. It turns into francium-221, then astatine-217, then bismuth-213, then mostly polonium-213, then lead-209 before finally hitting a hard stop at bismuth-209, which is stable. In most of these generations, the radiation released consists of alpha particles, which can destroy cancer cells but have low tissue penetration—they leave the surrounding healthy cells mostly alone. Currently, all but one of the radioactive isotopes used in cancer treatment release beta radiation, which causes considerably more collateral damage.

If a drug company could attach an atom of actinium-225 to a targeting system—like, say, the kind in CAR-T cells—the actinium-225 could continuously attack cancer for days at a time, like an artificial, radioactive version of the immune system. Newer chemotherapy drugs called antibody-drug conjugates already use this technique, directing chemotherapy agents that are too strong to give intravenously precisely where they are needed. At least two of these, Kadcyla and Adcetris, have already been approved by the FDA (for HER2-positive breast cancer and Hodgkin's lymphoma, respectively).

The U.S. system of national laboratories is already in talks with drug companies about making antibody-based radioactive drugs a reality. They seem promising: in a paper released last July in the *Journal of Nuclear Medicine,* one late-stage prostate cancer patient treated with three cycles of targeted actinium-225 at the University Hospital Heidelberg in Germany went into complete remission and another's tumors disappeared from scans.

But of course, there's a problem: now that the reactor program and the Cold War are both over, no one is making uranium-233 in the United States (or anywhere). And because it takes more than 40 years for uranium-233 to turn into enough thorium-229 to be useful, it wouldn't matter much even if they did. There are currently only about 1,500 to 1,700 millicuries of actinium-225 anywhere in the world, which would just treat 100 to 200 patients a year.

Which brings us to the reason Los Alamos has gotten deeply involved in actinium-225 at all: they're going to figure out how to make more from scratch.

Interlude

A roughshod man with bloodshot eyes rolls a cigarette outside a coffee shop in Taos, New Mexico. I can't be sure if he is the backpacker who was playing a flute at this table earlier or a new person. "You a reporter?" he asks.

"Er, yeah. Just got off the phone with a drug company that thinks they can cure cancer."

"A drug for cancer already exists," the man says. "More people need to be looking at marijuana. It can cure all kinds of sicknesses, but the thing is, the government doesn't want people knowing about it."

*A light breeze rustles the wind chimes. We are hiding from the sun under
a pergola on the shop's back porch. Another man attempts to come to my rescue: "But wasn't Obama trying to change the rules about experimenting—"*

*"Obama doesn't want to change the rules because he's not like us," says
the first man. "He's got pharaoh DNA that they blend with lizard blood up
in the mountains." He inclines his chin toward Los Alamos.*

"So he's like a monster?" asks the second man.

*"Nah, they're physical, like us, but they only have three chakras, so
they're not as balanced." He nods, sagely. "Highly carnivorous."*

V. Policy Reform—Or, Divided We Fall

Imagine cancer researchers as thousands of ships attempting to
cross the Pacific, all with skills and tools that they have perfected
in their home countries. Some have expert navigators. Others
build the most watertight ships. If someone could combine the
skills of the entire group, they could build a supership the likes
of which has never been seen. Instead, they seem to communicate
mostly by throwing paper airplanes at each other.

"All you could do with government-funded academic research,
in the age of paper, was share information in person, so you had
these huge cancer meetings once a year where everybody holds
their research until they get there," says Greg Simon, the executive
director of former vice president Joseph Biden's Cancer Moonshot, an initiative launched by the Obama administration in 2016.
"We haven't changed it since."

The system of medical journals, subscriptions to which can cost
thousands of dollars, are hardly the only baked-in obstacle to progress in cancer research. Clinical trials are still designed the same
way they were 50 years ago. Funding, applied for and received in
crazed round-robins of grant-writing, tends to reward low-risk experiments. There's secrecy and competition and slowness and inherent bureaucracy. The system wasn't created to be inefficient,
but now that it is, it is intractably so.

Just this week, Simon has flown all over the country trying to
bring bullheaded institutions with impossibly huge data troves
into a single kumbaya circle of progress. This morning, he gave a
speech at the 28th Annual Cancer Progress Conference. Now he is
entertaining a journalist at a sushi lunch in the lobby of a Manhat-

tan hotel. By rights, he should be asleep at the table with his face on a plate. Instead, he orders plain fish, no rice, in a disarming southern accent. (Simon is from Arkansas.)

When Simon was twenty-eight, he played drums in a rock band called the Great Zambini Brothers Band. Then he decided to do something with his life, "quit the band, waited tables, went to law school, got a job, and hated it," he says. A friend found him work in Washington and by forty-one, Simon was working in the White House as an aide to then vice president Al Gore. Then he cofounded a Washington think tank called FasterCures. Then he worked as senior vice president for patient engagement at Pfizer. If anyone on Earth knows how to get from here to there, Simon is the guy.

Since he left the White House (again) in January, Simon and his team have begun developing, out of a WeWork space, a spin off of the Cancer Moonshot they're calling the Biden Cancer Initiative. It will be its own separate nonprofit, apart from government or charity. Its goal: fix policy and make connections so that those with the expertise to cure cancer have a clear path to the finish line.

To achieve such a feat, Simon will work against a scientific version of the tragedy of the commons—an economic theory in which each person, acting in his own best interest, screws up the whole for everyone else. Convincing people and institutions to act against their own best interest will be much like governing, which is to say, slow and impossible. And yet it's hard not to believe in Biden, a man who helped run the most powerful country in the world at the same time he lost his own son to brain cancer.

"We won't be funding research. The world doesn't need another foundation with money," says Simon. "What it needs is someone like Biden, who's willing to knock heads together . . ." He pauses. "Or cajole heads together, to make the changes that everyone has an excuse not to do: I wanna make money, I want tenure, I wanna get published, I want this, I want that."

The fragmentation in medical research—the lone ships out on the ocean—doesn't exist as much in other sciences, says Simon, because scientists in other disciplines have no choice but to share equipment: telescopes or seismology sensors or space shuttles. Industries that have managed to work together have sent humans to the moon. "We don't even know how much progress we could make in our cancer enterprise because we've never had it up and running at a level that would be optimal," he says.

Simon himself had cancer. Three years ago. It was CLL. "I found it through a physical," he says. "I never had any of the raging symptoms, like bleeding. During the chemo I didn't notice it at all. Zero side effects. I thought I'd lose my hair so I grew a beard. But I didn't."

Interlude

"You are writing. Are you writer?" asks the flight attendant on Delta Flight 3866 from LaGuardia to Memphis in a thick Eastern European accent. It's a late flight—post-work—and many of the passengers are asleep. My reading light is one of just three that are illuminated.

"I had cancer," she says. "Breast cancer. I still have no boobs. After my surgery, they put in a balloon that they inflate step by step. After a few weeks I say to the doctor, 'I am still as flat as pancake!' And he says, 'Ah, there must be a hole.'"

The flight is turbulent, so the flight attendant perches on the arm of the seat in front of me. "I go home after surgery and I have a chill, so I take my medication—they give you such powerful medication—and I sleep. Thank god my friend came over and said I had to take a shower, because I took off the bandages and it was as red as this!" She points to the crimson bit of her Delta pin.

The flight attendant, diagnosed with stage-3a breast cancer, had developed a blood infection, and had to go to the hospital for intravenous antibiotics. After that, she had 8 rounds of chemotherapy and 33 of radiation.

"There was so much pain, but I had to walk through the pain. I made myself," she says. "I wrote 'I love you' on my mirror in lipstick. When you're single and you have cancer and you look at yourself, you need to read that. What else is there to do?"

VI. Silicon Valley—Or, The Brain

Through the floor-to-ceiling windows of the Parker Institute for Cancer Immunotherapy in San Francisco are the windswept headlands of the Golden Gate Bridge, the Pacific Ocean, and a frothy coral rotunda called the Palace of Fine Arts.

"Would you like a water?" asks the center's publicist when I visit. "Still or sparkling?"

Of all the cancer centers I visited, the Parker Institute seemed the most like the future of medicine. The office, a few doors from Lucasfilm, has one of those pristine, snack-filled tech startup kitchens with glass jars and a microwave that pulls out like an oven. On a table in the reception area sits a set of glittery silver pamphlets the size of small yearbooks explaining the mission.

The man behind the Parker Institute is serial entrepreneur Sean Parker, the cofounder of Napster and intermittent recipient of richly deserved tabloid jabs. Parker doesn't have the most sterling humanitarian reputation: In the movie *The Social Network*, Justin Timberlake portrayed him as a narcissistic party boy who screws over one of Facebook's cofounders and is arrested for cocaine possession. Parker was fined $2.5 million by the California Coastal Commission for building the set of his $10 million *Lord of the Rings*–themed wedding (complete with fake ruins, waterfalls, and a cottage) in an ecologically sensitive area. And yet, a little over a year ago, the same man donated $250 million to fund the study of immunotherapy at a lavish backyard gala featuring performances by John Legend, Lady Gaga, and the Red Hot Chili Peppers.

The public story about Parker's philanthropic effort is that it stemmed from the death of his close friend, film producer Laura Ziskin, to recurrent breast cancer. According to Jeff Bluestone, the Parker Institute's president and CEO (and, incidentally, the researcher who characterized CTLA-4 around the same time as Jim Allison), Parker was interested in cancer long before he met Ziskin. "Sean's been interested in the immune system for much of his life, because he's got asthma, and he's had a serious immunological imbalance," Bluestone says, sitting at a polished, raw-wood conference table half again as long as a normal conference table. (Parker is extremely allergic to peanuts.) "As long ago as 2004, before Laura got sick again, he thought the immune system was going to be the answer. He deeply understands a lot of the science. We joke, is he a second-year graduate student? A third-year postdoc? Should he just go get a Ph.D.?"

Parker is not the first very wealthy person who has used his money to combat disease. Several people at the institute took care to explain how they were different from the Howard Hughes

Medical Institute, a science-funding organization founded by the reclusive airman in 1953. A more influential predecessor might be Michael Milken, the Wall Street financier who founded a charity dedicated to family medicine with his brother Lowell in 1982 that supported, among other things, the research that led to Gleevec, the precision-medicine drug. Milken's funds also supported Jim Allison during an important time in his pre-checkpoint-inhibitor-therapy research when his National Institutes of Health grant had briefly lapsed. In 2003, Milken cofounded FasterCures with Greg Simon with the goal of increasing the pace of cures to "all serious diseases."

Some would argue that technology entrepreneurs are exactly the people who should be constructing the immaculate future of cancer research conceived by people like Joe Biden and Greg Simon. For one thing, tech entrepreneurs have already disrupted everything else. They understand the fast-moving, coin-chasing world of biotech development. Parker himself has already succeeded at convincing hardheaded institutions to work together. While he was an early investor and board member in the music streaming service Spotify, he negotiated with Universal and Warner to convince them to participate.

The Parker Institute's fundamental accomplishment thus far has been to do exactly this in cancer research. From the beginning, six academic research institutions signed on to work together under the Parker Institute's umbrella: Memorial Sloan Kettering; MD Anderson; Penn Medicine; Stanford Medicine; University of California, Los Angeles; and University of California, San Francisco. The six, along with independent investigators at a few other research institutions, agree to share research data and work together on goals and projects without getting hung up on institutional constraints, such as intellectual property. In return, they get two things: money, which every cancer researcher needs, and guidance, which is equally pressing but not necessarily as obvious.

"To become a leader in this field, to be a Carl June or a Jim Allison, you usually have to be a bit—not myopic, but a little blind," says Fred Ramsdell, the Parker Institute's vice president of research. This is common in science. To understand and work on a complicated concept, a researcher has to shut out the noise of everything except his exact area of expertise. Someone who works on checkpoint inhibitor therapies in melanoma, for example, might not see much use in reading about ovarian cancer detectors made out of nanocarbon—until suddenly it's the exact bridge to his own next level of progress.

"If a person knows nothing about nanoparticles, I can step in and say, Hey, this nanoparticle thing might be exactly what you need," says Ramsdell. "I spend a lot of time trying to develop relationships between people who might not always do so on their own." Some of those relationships are between researchers themselves. Others are between M.D.'s and Ph.D.'s, or between researchers and drug companies, or engineering companies, or the U.S. Patent Office. It doesn't really matter, so long as the arrangement furthers knowledge.

Up the coast in Seattle, another tech company is attempting to help cancer researchers cross entrenched divides. Microsoft's Project Hanover has already made considerable progress on creating a combined, searchable repository of the scientific news released every month by cancer researchers all over the world. The idea is that artificial intelligence will do a better job of parsing the vast landscape of scientific papers (those paper airplanes flying between ships) for insights. Rather than fallible humans trying to catch every valuable new detail as papers fly out of scientific clearinghouses, the system will do it for them, considering every possible combination of targeted drugs and genetic interactions in less time and more detail than it would take a team of educated humans.

Microsoft calls this the reasoning bottleneck. In a way, they're tackling it the same way the Parker Institute is. The same way the human body does: they're adding a brain.

Interlude

San Francisco. It's late. At the restaurant, there is a man seated at the chef's table when I arrive, drinking a balloon glass of red wine.

"How's the food?" the man asks after a good half hour. It is delicious —a buttery bucatini with lamb ragu and bread crumbs. The man has lived down the street from this restaurant for years. He's a former tech entrepreneur who is now a project manager for a retail company. I tell him what I am writing.

"That's a hell of a coincidence," the man says. "I just flew home from watching my father die of cancer."

"Jesus, I'm sorry."

"He's still there. With my sister. He told me he was tired of feeling like he was on death watch. He told me I should just go. So I went."

He sips his wine.

VII. Hope

What you see after a person has been debilitated by cancer and lived are the scars. The missing jaw or breast. The colostomy bag. Hair that has grown back curly or coarse or gray in patches. Tattoos that mark the paths of radiation beams. The disease that contains all of human biology leaves no one unchanged. There is before cancer, and then there is after.

Above Patrick Garvey's desk, on the top shelf of a bookcase, sits a stack of brown resin jawbones—dozens of them, mostly the mandible, or bottom jaw, which is commonly replaced with a bit of lower leg bone when it has to be removed because it is shot through with cancer. Every jawbone above Garvey's desk is a relic from a surgery he has performed at MD Anderson over the course of three years—more than 200 patients whose faces are forever altered by their interaction with the disease.

Later today, Garvey will operate on a man with a more difficult case—a large tumor in the maxilla, or top jaw—as part of two surgical teams. The first team will remove the tumor and most of the bone, including the man's eye, and then Garvey's team will remove a piece of the man's fibula along with its blood supply and use it to reconstruct the man's face. "We'll be here into the night," Garvey says.

This type of surgery is called microvascular reconstruction surgery. It drastically improves life for patients who would otherwise, like late film critic Roger Ebert, no longer be able to eat or talk without support. When it fails, however, it fails impressively: the transferred bone must have the correct blood supply or the body will simply reabsorb it, leaving only the bare metal scaffold the doctor implanted. Human bone is far better suited to the long-term mechanics of chewing and talking than metal is, and a plate without bone to protect it will eventually snap, like a paper clip bent back and forth over and over. Garvey has had to reconstruct jaws that have failed before, leaving patients disfigured and unable to chew properly. For a patient who has already undergone treatment for cancer, the impact of having to have multiple reconstructive face surgeries is harrowing.

To make the surgery simpler, Garvey's team uses 3-D-printed cutting guides and robotically milled metal plates to create the most precise reconstruction possible. This is how the brown resin jawbone graveyard above his desk got started. After a patient has a CT scan, a company called Materialise in Plymouth, Michigan, prints the jaw models as well as bolt-on cutting guides that show the surgeons exactly where to saw and reconnect fibula bones to match the person's original bone structure. Another company, in New York, creates a metal scaffold that is meticulously bent so as to recreate the original face angles, so MD Anderson's surgeons don't have to bend an off-the-shelf part into position during the reconstruction.

By all accounts, using 3-D-printed guides to reconstruct a human face is an advance at the very edge of cancer medicine, and yet it is *still* disheartening to look at the statistics. Last year, another 1.7 million Americans were diagnosed with cancer, and almost 600,000 died. Since 2004, according to the latest data available, the overall decline in death rates has been just 1.8 percent in men and 1.4 percent in women year over year. The five-year survival rate for pancreatic cancer, which most doctors consider the worst of the worst, sits stubbornly at just 8.2 percent.

Perhaps the cure for cancer seems so elusive because it's a failure of semantics. "Curing cancer" is impossible, and the statistics reflect that: cancer kills more Americans every two years than those who died in every war we've ever fought. However, helping some cancer patients, the luckiest of the unlucky, live in relative normalcy for years is not just possible. It is happening. The five-year overall cancer survival rate is up from 50 percent in 1975 to 67 percent today. For melanoma, it's 91.7 percent. For prostate cancer, it's 98.6. It will take time for the most promising treatments to trickle down to everyone they might be able to help, but in the meantime, the march continues.

What this has to do with Patrick Garvey is that, even subtly, using 3-D-cutting guides to improve plastic surgery shifts the focus of cancer treatment from emergency battlefield triage to matters of aesthetics and psychology that matter months and years down the line. Without saying it, exactly, the field of cancer treatment is acknowledging that cancer could one day become a survivable disease—that even a stage-4 metastatic cancer patient could survive long enough for normalcy to matter.

There are others on the frontlines: at hospitals across the country, women with breast cancer can wear a scalp-cooling system called DigniCap during chemotherapy treatments to reduce the likelihood of hair loss. At MD Anderson, a neuroscientist retrains patients' brains to improve altered nerve sensation caused by chemotherapy. St. Jude hired a psychologist to help teen cancer patients plan to save their eggs or sperm, in case their treatments render them infertile and they want to have a family in the future.

Future. A tricky word for a cancer patient. Who gets to have one is still a function of blind fortune. But all these ideas are starting to come together, and progress is suddenly accelerating. We are at what Crystal Mackall calls "the end of the beginning," and the hope is that one day soon, the miracles will no longer be miracles. They will just be what happens. Until then, we pin our hopes on the incremental or unpredictable improvements—the half measures, the better outcomes. It will always be true that once a person has had that most frightening of conversations with chance, life will be split into two parts—the time before cancer, and the time after it. But for a fortunate few, perhaps the second part can be as good, and even as rich and wonderful and great, as the first.

PART II

"What Do We Want? Evidence-Based Science. When Do We Want It? After Peer Review"

Rethinking Established Science

SOPHIE BRICKMAN

The Squeeze: Silicon Valley Reinvents the Breast Pump

FROM *California Sunday Magazine*

ON A RECENT Monday, four women gathered at a swanky coffee shop in San Francisco's SOMA neighborhood and politely asked an unshaven engineer to move down so they could access an outlet. After plugging in, they slipped what looked like portable mini mist fans under their shirts.

If the engineer had been paying attention—instead of taking a break from his coding to scroll through a Wikipedia page on the head coach of the Houston Rockets—he might have been able to make out the wheezing sound of breast pumps at work, or noticed that four nearby bustlines had each grown an inch.

"I work, don't have an office, have a seven-week-old at home, and don't want to pump in the bathroom," said Chloe Sladden, former VP of media at Twitter and a founding investor with #Angels, a women-led venture capital group. Amid ubiquitous laptop shoulder bags sat her tote, decorated with floating pairs of painted boobs.

The other women, who had gathered to test-drive the latest product from a startup called Moxxly, nodded vigorously. Instead of having to take off their shirts and affix two *Xena: Warrior Princess*–style cones onto their breasts, the Moxxly Flow, pump attachments set to hit the market this fall, allowed for discreet, under-the-bra, hands-free pumping. Two of the pumpers, including Sladden, are also investors.

Sladden adjusted her silk shirt, which draped loosely over her newly enhanced bosom. "It's about time we focused on this part of motherhood," she said.

The first mechanical breast pump, modeled after bovine milking machines, was invented in the 1920s, and little has changed about the fundamental design. A cone—or flange, as it's called—is placed on each breast. A collection bottle is screwed into the bottom of each flange and dangles off the boob like a particularly unflattering nipple tassel. Long plastic tubes connect the flanges to a mechanical pump that suctions out the milk. What this means in practice: women have to find a private corner, remove their shirts, and watch their nipples get rhythmically suctioned for 20 minutes until they're fully milked. If Mom would like to do something with her hands besides hold the cones in place, she needs to wear a sort of medieval corset, i.e., a tight-fitting pumping bra.

"It's just undignified and unjust," says Cara Delzer, CEO and cofounder of Moxxly.

Delzer, who graduated from Stanford's business school, was working in product marketing at eBay when she had her first child in 2013. "I'd be in a meeting, then 30 seconds later find myself with my shirt off," she recalls. "I had a really supportive team, I had pumping rooms, the best setup anyone could have, but still it was untenable." To men who might not understand the situation, she likens it to having to take your pants off for a half hour before hopping into a boardroom to deliver an important presentation.

But the logistics were just one issue. Another was packaging. The free pump her insurance company provided arrived in a cardboard box with a dry user manual. "It made me feel like I was a patient, not a woman," Delzer remembers. She wandered the aisles in baby stores and found countless products marketed with the baby in mind. Why wasn't there a brand out there that appealed to her?

In 2014, she partnered with two other Stanford graduates—designer Gabrielle Guthrie, who'd reimagined the breast pump for her master's thesis, and Santhi Analytis, a mechanical engineer who worked with Guthrie on a breast pump redesign at a hackathon. They were accepted to Highway1, a hardware startup accelerator in San Francisco, where they began developing the prototype for what would become the Moxxly Flow: first, duct-taping existing parts together, then moving to 3-D printing. A timeline of the Flow's evolution looks a bit like the *March of Progress*. With each iteration, the bottle gets more upright, the flange a little prouder.

Then one of the greatest challenges came to a head: raising capital. According to Transparency Market Research, the breast pump market is set to reach $2.6 billion worldwide by 2020, but as Delzer learned, "VC is a game of pattern recognition. And there's no pattern here." She'd often find herself in pitch meetings slipping parts under her bra and awkwardly explaining to a room full of VC partners—93 percent of whom are men—the indignities of the pumping life.

After dozens of pitch meetings, she managed to secure enough funding from well-known investors like Slow Ventures (which counts Evernote, Dropbox, and Pinterest in its portfolio) and Scott and Cyan Banister (the husband-and-wife team that sits on the boards of PayPal and Postmates) to bring Moxxly to market. "These other large [pump] companies have this relationship with insurance and health providers that is baked in, and they're not incentivized to be more human-centered or focused on the modern woman," says Enrique Allen of Designer Fund, which invested in an early round. "We often look for market opportunity where there are big, sleepy incumbents who don't have a mobile strategy or a way to reach a new audience in the way that Moxxly does."

But, Allen admits, his lactation learning curve was steep. "I didn't realize nipples could get so sore," he says. "I definitely have more empathy now."

Other companies have seen the market opening as well, and Silicon Valley is finally—in industry parlance—disrupting the hell out of the space. The Willow, set to come out later this year, is a wearable pump that fits entirely inside a bra. It looks so sleek and Jobsian, you'd almost expect it to play Drake when you press the power button. The Naya pump uses water-based suction instead of air, which they claim is quieter and more comfortable. Babyation, coming at the end of this year, is designed so the collection bottles don't hang off the breasts.

For all their improvements, however, the new pumps don't come cheap. At under $100, the Moxxly Flow attachment is one of the least expensive options. The Naya retails for a whopping $999, while the Willow, whose price is still being decided, is in the ballpark of $400 for the pump, which doesn't include the cost of each single-use collection bag. So until insurance companies embrace pump 2.0, many women will have to stick with the free, bovine-esque milkers.

The Moxxly offices are located in an old factory in the Bayview neighborhood of San Francisco. Next to a 3-D printer and some special rulers that measure nipple and areola size is a cardboard box of "old test boobs." It's the domain of Jake Kurzrock, a mechanical engineer with a handlebar mustache who, in a past life, worked in fungal genes and fertilizer.

"First, we were using a stress boob for testing," he said, pulling out a novelty stress ball colored to look like a Caucasian breast, with a baby bottle nipple crammed in the middle. Next to it sat an African American companion stress breast. "We tried these cutlets, too," he said, slapping some mangled silicone bra inserts onto a worktable, invoking a Hannibal Lecter–ish experiment gone wrong. "Then I just went on Amazon and got a breast used by cross-dressers. You can get really nice ones, like hundreds of dollars. We got the $40 one. It worked fine." He used it to make a mold for breasts they'd eventually affix to a dress form named Seraphina.

Several factors come into play when constructing a piece of hardware that needs to work across varying body types. There's the fact that nipple shape and size vary by ethnicity (cocktail party factoid: Asian nipples tend to be longer). Plus, the rate and angle at which milk flows through the breasts differ for every individual (the Moxxly team found that, for modeling purposes, warm skim best mimics human). All calculations then have to account for, as one Google spreadsheet put it, "possible sagginess." The result of all this research? "It makes boobs less fun," Kurzrock said.

The women at the coffee shop, who'd collectively pumped enough ultralocal milk for several lattes, would likely disagree. As Sladden threw her bottles into her boob-print tote, one of the other women recalled the day she unthinkingly wore a shift dress and was forced to get virtually naked for her pumping sessions. Another remembered being on a work camping trip, topless and sliding down in her car's front seat, away from the prying eyes of coworkers who meandered outside. Those humiliating pumping days—for this particular set of women, at least—seemed to be behind them.

JOHN LANCHESTER

The Case Against Civilization

FROM *The New Yorker*

SCIENCE AND TECHNOLOGY, we tend to think of them as siblings, perhaps even as twins, as parts of STEM (for "science, technology, engineering, and mathematics"). When it comes to the
shiniest wonders of the modern world—as the supercomputers in
our pockets communicate with satellites—science and technology
are indeed hand in glove. For much of human history, though,
technology had nothing to do with science. Many of our most significant inventions are pure tools, with no scientific method behind them. Wheels and wells, cranks and mills and gears and ships'
masts, clocks and rudders and crop rotation: all have been crucial
to human and economic development, and none historically had
any connection with what we think of today as science. Some of
the most important things we use every day were invented long
before the adoption of the scientific method. I love my laptop and
my iPhone and my Echo and my GPS, but the piece of technology
I would be most reluctant to give up, the one that changed my life
from the first day I used it, and that I'm still reliant on every waking hour—am reliant on right now, as I sit typing—dates from the
thirteenth century: my glasses. Soap prevented more deaths than
penicillin. That's technology, not science.

In *Against the Grain: A Deep History of the Earliest States,* James C.
Scott, a professor of political science at Yale, presents a plausible
contender for the most important piece of technology in the history of man. It is a technology so old that it predates *Homo sapiens*
and instead should be credited to our ancestor *Homo erectus.* That
technology is fire. We have used it in two crucial, defining ways.

The first and the most obvious of these is cooking. As Richard Wrangham has argued in his book *Catching Fire,* our ability to cook allows us to extract more energy from the food we eat, and also to eat a far wider range of foods. Our closest animal relative, the chimpanzee, has a colon three times as large as ours, because its diet of raw food is so much harder to digest. The extra caloric value we get from cooked food allowed us to develop our big brains, which absorb roughly a fifth of the energy we consume, as opposed to less than a tenth for most mammals' brains. That difference is what has made us the dominant species on the planet.

The other reason fire was central to our history is less obvious to contemporary eyes: we used it to adapt the landscape around us to our purposes. Hunter-gatherers would set fires as they moved, to clear terrain and make it ready for fast-growing, prey-attracting new plants. They would also drive animals with fire. They used this technology so much that, Scott thinks, we should date the human-dominated phase of Earth, the so-called Anthropocene, from the time our forebears mastered this new tool.

We don't give the technology of fire enough credit, Scott suggests, because we don't give our ancestors much credit for their ingenuity over the long period—95 percent of human history—during which most of our species were hunter-gatherers. "Why human fire as landscape architecture doesn't register as it ought to in our historical accounts is perhaps that its effects were spread over hundreds of millennia and were accomplished by 'precivilized' peoples also known as 'savages,'" Scott writes. To demonstrate the significance of fire, he points to what we've found in certain caves in southern Africa. The earliest, oldest strata of the caves contain whole skeletons of carnivores and many chewed-up bone fragments of the things they were eating, including us. Then comes the layer from when we discovered fire, and ownership of the caves switches: the human skeletons are whole, and the carnivores are bone fragments. Fire is the difference between eating lunch and being lunch.

Anatomically modern humans have been around for roughly 200,000 years. For most of that time, we lived as hunter-gatherers. Then, about 12,000 years ago, came what is generally agreed to be the definitive before-and-after moment in our ascent to planetary dominance: the Neolithic Revolution. This was our adoption of, to use Scott's word, a "package" of agricultural innovations, notably the domestication of animals such as the cow and the pig, and the transition from hunting and gathering to planting and cultivating crops. The most important of these crops have been the cereals —wheat, barley, rice, and maize—that remain the staples of humanity's diet. Cereals allowed population growth and the birth of cities, and, hence, the development of states and the rise of complex societies.

The story told in *Against the Grain* heavily revises this widely held account. Scott's specialty is not early human history. His work has focused on a skeptical, peasant's-eye view of state formation; the trajectory of his interests can be traced in the titles of his books, from *The Moral Economy of the Peasant* to *The Art of Not Being Governed.* His best-known book, *Seeing Like a State,* has become a touchstone for political scientists, and amounts to a blistering critique of central planning and "high modernism," the idea that officials at the center of a state know better than the people they are governing. Scott argues that a state's interests and the interests of subjects are often not just different but opposite. Stalin's project of farm collectivization "served well enough as a means whereby the state could determine cropping patterns, fix real rural wages, appropriate a large share of whatever grain was produced, and politically emasculate the countryside"; it also killed many millions of peasants.

Scott's new book extends these ideas into the deep past, and draws on existing research to argue that ours is not a story of linear progress, that the timeline is much more complicated, and that the causal sequences of the standard version are wrong. He focuses his account on Mesopotamia—roughly speaking, modern-day Iraq—because it is "the heartland of the first 'pristine' states in the world," the term *pristine* here meaning that these states bore no watermark from earlier settlements and were the first time any such social organizations had existed. They were the first states to have written records, and they became a template for other states in the Near East and in Egypt, making them doubly relevant to later history.

The big news to emerge from recent archeological research concerns the time lag between "sedentism," or living in settled communities, and the adoption of agriculture. Previous scholarship held that the invention of agriculture made sedentism possible. The evidence shows that this isn't true: there's an enormous gap—4,000 years—separating the "two key domestications," of animals and cereals, from the first agrarian economies based on them. Our ancestors evidently took a good, hard look at the possibility of agriculture before deciding to adopt this new way of life. They were able to think it over for so long because the life they lived was remarkably abundant. Like the early civilization of China in the Yellow River Valley, Mesopotamia was a wetland territory, as its name ("between the rivers") suggests. In the Neolithic period, Mesopotamia was a delta wetland, where the sea came many miles inland from its current shore.

This was a generous landscape for humans, offering fish and the animals that preyed on them, fertile soil left behind by regular flooding, migratory birds, and migratory prey traveling near river routes. The first settled communities were established here because the land offered such a diverse web of food sources. If one year a food source failed, another would still be present. The archeology shows, then, that the "Neolithic package" of domestication and agriculture did not lead to settled communities, the ancestors of our modern towns and cities and states. Those communities had been around for thousands of years, living in the bountiful conditions of the wetlands, before humanity committed to intensive agriculture. Reliance on a single, densely planted cereal crop was much riskier, and it's no wonder people took a few millennia to make the change.

So why did our ancestors switch from this complex web of food supplies to the concentrated production of single crops? We don't know, although Scott speculates that climatic stress may have been involved. Two things, however, are clear. The first is that, for thousands of years, the agricultural revolution was, for most of the people living through it, a disaster. The fossil record shows that life for agriculturalists was harder than it had been for hunter-gatherers. Their bones show evidence of dietary stress: they were shorter, they were sicker, their mortality rates were higher. Living in close proximity to domesticated animals led to diseases that crossed the species barrier, wreaking havoc in the densely settled communities. Scott calls them not towns but "late-Neolithic multi-species resettlement camps." Who would choose to live in one of those? Jared Diamond called the Neolithic Revolution "the worst mistake in human history." The startling thing about this claim is that, among historians of the era, it isn't very controversial.

The other conclusion we can draw from the evidence, Scott says, is that there is a crucial, direct link between the cultivation of cereal crops and the birth of the first states. It's not that cereal grains were humankind's only staples; it's just that they were the only ones that encouraged the formation of states. "History records no cassava states, no sago, yam, taro, plantain, breadfruit or sweet potato states," he writes. What was so special about grains? The answer will make sense to anyone who has ever filled out a Form 1040: grain, unlike other crops, is easy to tax. Some crops (potatoes, sweet potatoes, cassava) are buried and so can be hidden from the tax collector, and, even if discovered, they must be dug up individually and laboriously. Other crops (notably, legumes) ripen at different intervals, or yield harvests throughout a growing season rather than along a fixed trajectory of unripe to ripe —in other words, the taxman can't come once and get his proper due. Only grains are, in Scott's words, "visible, divisible, assessable, storable, transportable, and 'rationable.'" Other crops have some of these advantages, but only cereal grains have them all, and so grain became "the main food starch, the unit of taxation in kind, and the basis for a hegemonic agrarian calendar." The taxman can come, assess the fields, set a level of tax, then come back and make sure he's got his share of the harvest.

It was the ability to tax and to extract a surplus from the produce of agriculture that, in Scott's account, led to the birth of the state, and also to the creation of complex societies with hierarchies, division of labor, specialist jobs (soldier, priest, servant, administrator), and an elite presiding over them. Because the new states required huge amounts of manual work to irrigate the cereal crops, they also required forms of forced labor, including slavery; because the easiest way to find slaves was to capture them, the states had a new propensity for waging war. Some of the earliest images in human history, from the first Mesopotamian states, are of slaves being marched along in neck shackles. Add this to the frequent epidemics and the general ill health of early settled communities and it is not hard to see why the latest consensus is that the Neolithic Revolution was a disaster for most of the people who lived through it.

War, slavery, rule by elites—all were made easier by another new technology of control: writing. "It is virtually impossible to conceive of even the earliest states without a systematic technology of numerical record keeping," Scott maintains. All the good things we associate with writing—its use for culture and entertainment and communication and collective memory—were some distance in the future. For half a thousand years after its invention, in Mesopotamia, writing was used exclusively for bookkeeping: "the massive effort through a system of notation to make a society, its manpower, and its production legible to its rulers and temple officials, and to extract grain and labor from it." Early tablets consist of "lists, lists, and lists," Scott says, and the subjects of that record-keeping are, in order of frequency, "barley (as rations and taxes), war captives, male and female slaves." Walter Benjamin, the great German Jewish cultural critic, who committed suicide while trying to escape Nazi-controlled Europe, said that "there is no document of civilization which is not at the same time a document of barbarism." He meant that every complicated and beautiful thing humanity ever made has, if you look at it long enough, a shadow, a history of oppression. As a matter of plain historical fact, that seems right. It was a long and traumatic journey from the invention of writing to your book club's discussion of Jodi Picoult's latest.

We need to rethink, accordingly, what we mean when we talk about ancient "dark ages." Scott's question is trenchant: "'dark' for whom and in what respects"? The historical record shows that early cities and states were prone to sudden implosion. "Over the roughly five millennia of sporadic sedentism before states (seven millennia if we include preagriculture sedentism in Japan and the Ukraine)," he writes, "archaeologists have recorded hundreds of locations that were settled, then abandoned, perhaps resettled, and then again abandoned." These events are usually spoken of as "collapses," but Scott invites us to scrutinize that term, too. When states collapse, fancy buildings stop being built, the elites no longer run things, written records stop being kept, and the mass of the population goes to live somewhere else. Is that a collapse, in terms of living standards, for most people? Human beings mainly lived outside the purview of states until—by Scott's reckoning—about the year 1600 A.D. Until that date, marking the last two-tenths of one percent of humanity's political life, "much of the world's population might never have met that hallmark of the state: a tax collector."

The question of what it was like to live outside the settled culture of a state is therefore an important one for the overall assessment of human history. If that life was, as Thomas Hobbes described it, "nasty, brutish, and short," this is a vital piece of information for drawing up the account of how we got to be who we are. In essence, human history would become a straightforward story of progress: most of us were miserable most of the time, we developed civilization, everything got better. If most of us *weren't* miserable most of the time, the arrival of civilization is a more ambiguous event. In one column of the ledger, we would have the development of a complex material culture permitting the glories of modern science and medicine and the accumulated wonders of art. In the other column, we would have the less good stuff, such as plague, war, slavery, social stratification, rule by mercilessly appropriating elites, and Simon Cowell.

To know what it is like to live as people lived for most of human history, you would have to find one of the places where traditional hunting-and-gathering practices are still alive. You would have to spend a lot of time there, to make sure that what you were seeing wasn't just a snapshot, and that you had a real sense of the texture of lived experience; and, ideally, you would need a point of comparison, people with close similarities to your hunter-gatherers but who lived differently, so that you would have a scientific "control" that allowed you to rule out local accidents of circumstance. Fortunately for us, the anthropologist James Suzman did exactly that: he spent more than two decades visiting, studying, and living among the Bushmen of the Kalahari, in southwest Africa. It's a story he recounts in his new book, *Affluence Without Abundance: The Disappearing World of the Bushmen.*

The Bushmen have long been of interest to anthropologists and scientists. About 150,000 years ago, 50,000 years after the emergence of the first anatomically modern humans, one group of *Homo sapiens* was living in southern Africa. The Bushmen, or Khoisan, are still there: the oldest growth on the human family tree. (The term "Bushman," once derogatory, is now used by the people themselves, and by NGOs, "invoking as it does a set of positive if romantic stereotypes," Suzman notes, though some Khoisan prefer to use the term "San.") The genetic evidence suggests that, for much of that 150,000 years, they were the largest population of biologically modern humans. Their languages use palatal clicks, such as a *tsk*, made by bringing the tongue back from the front teeth while gently sucking in air, and the "click" we make by pushing the tongue against the roof of the mouth, then bringing it suddenly downward. This raises the fascinating possibility that click languages are the oldest surviving variety of speech.

Suzman first visited the Bushmen in 1992, and went to stay with them two years later, as part of the research for his Ph.D. The group he knows best are the Ju/'hoansi, between 8,000 and 10,000 of whom are alive today, occupying the borderlands between Namibia and Botswana. (The phonetic mark /' represents a *tsk*.) The Ju/'hoansi are about 10 percent of the total Bushman population in southern Africa, and they are divided into a northern group, who retain significant control over their traditional lands, and who therefore still have the ability to practice hunting and gathering, and a southern group, who were deprived of their lands and "resettled" into modern ways of living.

To a remarkable extent, Suzman's study of the Bushmen supports the ideas of *Against the Grain*. The encounter with modernity has been disastrous for the Bushmen: Suzman's portrait of the dispossessed, alienated, suffering Ju/'hoansi in their miserable resettlement camps makes that clear. The two books even confirm each other's account of that sinister new technology called writing. Suzman's Bushman mentor, !A/ae, "noted that whenever he started work at any new farm, his name would be entered into an employment ledger, documents that over the decades had assumed great mystical power among Ju/'hoansi on the farms. The secrets held by these ledgers evidently had the power to give or withhold pay, issue rations, and determine an individual's right to stay on any particular farm."

It turns out that hunting and gathering is a good way to live. A study from 1966 found that it took a Ju/'hoansi only about 17 hours a week, on average, to find an adequate supply of food; another 19 hours were spent on domestic activities and chores. The average caloric intake of the hunter-gatherers was 2,300 a day, close to the recommended amount. At the time these figures were first established, a comparable week in the United States involved

40 hours of work and 36 of domestic labor. Ju/'hoansi do not accumulate surpluses; they get all the food they need, and then stop. They exhibit what Suzman calls "an unyielding confidence" that their environment will provide for their needs.

The web of food sources that the hunting-and-gathering Ju/'hoansi use is, exactly as Scott argues for Neolithic people, a complex one, with a wide range of animal protein, including porcupines, kudu, wildebeests, and elephants, and 125 edible plant species, with different seasonal cycles, ecological niches, and responses to weather fluctuations. Hunter-gatherers need not only an unwritten almanac of dietary knowledge but what Scott calls a "library of almanacs." As he suggests, the step-down in complexity between hunting and gathering and domesticated agriculture is as big as the step-down between domesticated agriculture and routine assembly work on a production line.

The news here is that the lives of most of our progenitors were better than we think. We're flattering ourselves by believing that their existence was so grim and that our modern, civilized one is, by comparison, so great. Still, we are where we are, and we live the way we live, and it's possible to wonder whether any of this illuminating knowledge about our hunter-gatherer ancestors can be useful to us. Suzman wonders the same thing. He discusses John Maynard Keynes's famous 1930 essay "The Economic Possibilities for Our Grandchildren." Keynes speculated that if the world continued to get richer we would naturally end up enjoying a high standard of living while doing much less work. He thought that "the economic problem" of having enough to live on would be solved, and "the struggle for subsistence" would be over:

> When the accumulation of wealth is no longer of high social importance, there will be great changes in the code of morals. We shall be able to rid ourselves of many of the pseudo-moral principles which have hag-ridden us for two hundred years, by which we have exalted some of the most distasteful of human qualities into the position of the highest virtues. We shall be able to afford to dare to assess the money-motive at its true value. The love of money as a possession — as distinguished from the love of money as a means to the enjoyments and realities of life — will be recognized for what it is, a somewhat disgusting morbidity, one of those semi-criminal, semi-pathological propensities which one hands over with a shudder to the specialists in mental disease.

The world has indeed got richer, but any such shift in morals and values is hard to detect. Money and the value system around its acquisition are fully intact. Greed is still good.

The study of hunter-gatherers, who live for the day and do not accumulate surpluses, shows that humanity can live more or less as Keynes suggests. It's just that we're choosing not to. A key to that lost or forsworn ability, Suzman suggests, lies in the ferocious egalitarianism of hunter-gatherers. For example, the most valuable thing a hunter can do is come back with meat. Unlike gathered plants, whose proceeds are "not subject to any strict conventions on sharing," hunted meat is very carefully distributed according to protocol, and the people who eat the meat that is given to them go to great trouble to be rude about it. This ritual is called "insulting the meat," and it is designed to make sure the hunter doesn't get above himself and start thinking that he's better than anyone else. "When a young man kills much meat," a Bushman told the anthropologist Richard B. Lee, "he comes to think of himself as a chief or a big man, and he thinks of the rest of us as his servants or inferiors . . . We can't accept this." The insults are designed to "cool his heart and make him gentle." For these hunter-gatherers, Suzman writes, "the sum of individual self-interest and the jealousy that policed it was a fiercely egalitarian society where profitable exchange, hierarchy, and significant material inequality were not tolerated."

This egalitarian impulse, Suzman suggests, is central to the hunter-gatherer's ability to live a life that is, on its own terms, affluent, but without abundance, without excess, and without competitive acquisition. The secret ingredient seems to be the positive harnessing of the general human impulse to envy. As he says, "If this kind of egalitarianism is a precondition for us to embrace a post-labor world, then I suspect it may prove a very hard nut to crack." There's a lot that we could learn from the oldest extant branch of humanity, but that doesn't mean we're going to put the knowledge into effect. A socially positive use of envy—now, that would be a technology almost as useful as fire.

SIDDHARTHA MUKHERJEE

Cancer's Invasion Equation

FROM *The New Yorker*

OVER THE SUMMER of 2011, the water in Lake Michigan turned crystal clear. Shafts of angled light lit the lake bed, like searchlights from a UFO; later, old sunken ships came into view from above. Pleasure was soon replaced by panic: lakes are not supposed to look like swimming pools. When biologists investigated, they found that the turbid swirls of plankton that typically grow in the lake by the million had nearly vanished—consumed gradually, they could only guess, by some ravenous organism.

The likely culprits were mollusks: the zebra mussel and its cousin the quagga mussel. The two species—*Dreissena polymorpha* and *Dreissena bugensis*—are thought to have originated in the estuarine basins of Ukraine, notably that of the Dnieper River. In the late 1980s, cargo ships, traveling from the Caspian Sea and the Black Sea, had dumped their ballast water into the Great Lakes, contaminating them with foreign organisms.

At first, the mollusks seemed like relatively innocuous guests. Then things took a turn. By the mid-1990s, they were hanging from ship keels, turbines, and propellers in bulbous, tumorlike masses, encrusting docks and piers, clogging water pipes and sanitation systems, and washing ashore in such numbers that, on some beaches, you could walk on a solid bar of shells. Eventually, the water clarity began to increase, the effect at first picturesque and then eerie.

By 2012, the *Dreissena* population in parts of southern Lake Michigan had reached a density of 10,000 per square meter. By one estimate, there were 950 *trillion* mussels in the lake, its bottom a crackling carpet of calcium. By 2015, the density was 15,000 per square meter—more mussels, by weight, than all the fish in the lakes. Billions of dollars in damage had accumulated. Ships and boats had to be decontaminated, and water-cleaning equipment dismantled and stripped. Dire warning signs ("DON'T MOVE A MUSSEL!") were placed throughout the lake system, yet the invaders—the quaggas, ultimately, in the greatest numbers—continued to spread.

What made the mussels such malignant invaders? Some of their aggression is a feature of their biology. The *Dreissena* are champion breeders, each churning out more than a million eggs a year. Yet in the basins and the deltas of Ukraine these mussels seldom reach even a fifth of their peak density in the Great Lakes. They rarely invade depths below 30 meters, clump on boats, clog marine equipment, or form calcified masses. They are, in short, a relatively docile species—restricted, perhaps, by the quality of the water, by their natural predators and pathogens, by the shallowness of the river basin, or by factors we haven't yet identified.

Solving the quagga conundrum requires cracking two halves of a puzzle. Half the story lies in the mussel's intrinsic biology—its genes, its morphology, its nutritional preferences, its reproductive habits. The other half involves the match between that biology and the environment. It is a basic insight that an undergraduate ecologist might find familiar: the "invasiveness" of an organism is always a relative concept. The Asian carp—another fierce aggressor in American waters—is not particularly invasive in parts of Asia. The Japanese knotweed, now colonizing the cherished gardens of the English, is hardly known as a weed in Japan. An aggressor in one environment is a placid resident in another. The meek are only circumstantially meek; when conditions change, they might suddenly inherit the Earth.

One evening this past June, as I walked along the shore of Lake Michigan in Chicago, I thought about mussels, knotweed, and cancer. Tens of thousands of people had descended on the city to attend the annual meeting of the American Society of Clinical Oncology, the world's preeminent conference on cancer. Much of the meeting, I knew, would focus on the intrinsic properties of cancer cells, and on ways of targeting them. Yet those properties might be only part of the picture. We want to know which mollusk we're dealing with, but we also need to know which lake.

A few weeks before the ASCO meeting, at Columbia University's hospital on 168th Street, I met a woman with breast cancer. Anna Guzello, a supermarket cashier from Brooklyn, had noticed a small lump in her left breast a few months earlier. (I've changed some of her identifying details.) A mammogram then revealed a hazy, spidery mass, and a biopsy confirmed that the tumor was malignant.

Guzello had a total mastectomy of the breast—a simple lumpectomy would not have sufficed, given the size and the location of the mass—and planned to have surgical reconstruction. On an afternoon in May, she came to see Katherine Crew, a breast oncologist at Columbia, to discuss the next steps in her treatment.

Crew's office, on the 10th floor of the hospital, is a small, square, sparsely furnished room. The light from a fluorescent desk lamp was flickering, and Crew switched it off. She wanted no distractions. Guzello, her hair coiled into a tight bun, leaned forward, frowning intently, as Crew drew pictures and wrote notes on a sheet of paper.

"Can you read my writing?" Crew asked. "You can keep the notes and always come back with questions." Her tone was gentle, but it was as if the weight of every word were multiplied.

Guzello nodded. She drummed her fingernails on the table, producing a staccato, military sound—*click-click-click*—a nervous tic that seemed to calm her.

"First, the good news," Crew said. "There's no visible cancer left in your body."

The surgeons had removed the tumor, with wide margins on all sides. The lymph nodes in the armpits—a frequent site of cancer metastasis—also contained no sign of cancer. In oncology parlance, Guzello would be classified as NED: "no evidence of disease."

But that's a squirrelly phrase: "evidence" refers to the state of our knowledge, not the state of the disease. Breast-cancer cells could have escaped and settled in Guzello's brain, spinal cord, or bones, where they might be invisible to scans and tests. Women with complete mastectomies and "no evidence of disease" can relapse with metastatic breast cancer months, years, or even decades after the removal of the primary cancerous mass. Patients who succumb to cancer generally die of these metastases, not of their primary tumors. (Notable exceptions are brain cancers, which can kill patients by occupying the skull, and blood cancers, in which the cancerous cells are inherently metastatic.)

"So we treat with medicines to decrease the chance of metastasis — the growth of cancer cells in sites outside the breast," Crew told Guzello. She explained that the medicines came in three main categories: cell-killing chemotherapy; targeted therapies, like Herceptin, that specifically go after the products of misbehaving genes in cancer cells; and estrogen-blocking pills, which are typically prescribed for five or ten years.

Guzello moved her hands over her hair, her lips tightening. The hormonal pills were fine. But she balked at the cell-killing chemotherapy.

"If I don't have those metastases, then I'll be taking risks for no reason," she said. The nails drummed on the table again. The risks were substantial: hair loss, diarrhea, infections, a small possibility of permanent numbness that would leave her hands feeling as if she were wearing leather gloves, yet exquisitely sensitive to cold. The chemotherapy protocol meant that she would be yoked to an IV pole at an infusion center for several hours once a week, for nearly half a year. She had a mother with a severe disability to care for, and few vacation days. Was there any way to know whether she was likely to suffer metastasis? "Then I'd be able to assess the risks and benefits more realistically," Guzello said.

The question has echoed through oncology for decades. We aren't particularly adept at predicting whether a specific patient's cancer will become metastatic or not. Metastasis can seem "like a random act of violence," Daniel Hayes, a breast oncologist at the University of Michigan, told me when we spoke at the ASCO meeting in Chicago. "Because we're not very good at telling whether breast-cancer patients will have metastasis, we tend to treat them with chemotherapy as if they all have potential metastasis." Only some fraction of patients who receive toxic chemotherapy will really benefit from it, but we don't know which fraction. And so, unable to say whether any particular patient will benefit, we have no choice but to overtreat. For women like Guzello, then, the central puzzle is not the perennial "why me." It's "whether me."

*

There are deep roots to the idea that a cancer's metastases depend on local habitats. In 1889, an English doctor named Stephen Paget set out to understand cancer's "primary growth and the situation of the secondary growths derived from it." The son and nephew of prominent English doctors—his father, James Paget, was one of the founders of modern pathology; his uncle was a Cambridge professor of medicine—the younger Paget might have been burdened by the deadweight of inherited wisdom. Cancer, in Paget's time, was thought to diffuse from its primary site like a malignant inkblot. Surgeons, believing this "centrifugal theory"—cancer's stainlike, outward spread from a central mass—advocated ever-widening surgical extirpations to eliminate cancer. (This theory would form the intellectual basis for William Halsted's "radical" mastectomy.) But when Paget collected the case files of 735 women who had died of breast cancer, he found a bizarre pattern of metastatic spread. The metastases didn't appear to spread centrifugally; they appeared in discrete, anatomically distant sites. And the pattern of spread was far from random: cancers had a strange and strong preference for particular organs. Of the 300-odd metastases, Paget found 241 in the liver, 17 in the spleen, and 70 in the lungs. Enormous, empty, uncolonized steppes—anatomical landmasses untouched by metastasis—stretched out in between.

Why was the liver so hospitable to metastasis, while the spleen, which had similarities in blood supply, size, and proximity, seemed relatively resistant? As Paget probed deeper, he found that cancerous growth even favored particular sites within organ systems. Bones were a frequent site of metastasis in breast cancer—but not every bone was equally susceptible. "Who has ever seen the bones of the hands or the feet attacked by secondary cancer?" he asked. Paget coined the phrase "seed and soil" to describe the phenomenon. The seed was the cancer cell; the soil was the local ecosystem where it flourished, or failed to. Paget's study concentrated on patterns of metastasis within a person's body. The propensity of one organ to become colonized while another was spared seemed to depend on the nature or the location of the organ—on local ecologies. Yet the logic of the seed-and-soil model ultimately raises the question of global ecologies: why does one person's body have susceptible niches and not another's?

Paget's way of framing the issue—metastasis as the result of a pathological relationship between a cancer cell and its environment—lay dormant for more than a century. There were exceptions. The pioneering metastasis researcher Isaiah J. Fidler, working at the National Cancer Institute during the 1970s and 1980s, started to study "cross-talk" between tissue and tumor. A tumor, Fidler showed, is made of a heterogeneous mixture of millions of cells, only a fraction of which are equipped to leave the primary tumor, form an exploitative alliance with the "soil" of another organ, and initiate metastasis. In the same period, Mina Bissell, working at the University of California, Berkeley, and then at the Lawrence Berkeley National Laboratory, began scrutinizing the microenvironments in which tumors formed—or didn't—as she looked for factors that enabled or disabled the growth of cancer in various organs. Context, she found, was critical.

Yet oncology as a whole remained dominated by a simpler model. When I was a medical student in Boston, I spent an evening in a frigid deli on Boylston Street memorizing the list of bone-metastasizing cancers (breast, lung, thyroid, kidney, prostate) using the unsavory mnemonic "BLT with kosher pickle" and coming up with a mental image of how metastases might form. Cancer "disseminated" via blood vessels, "attacked" the organs, and began to sprout and flourish there. As I rotated through the cancer wards in the late 1990s, doctors reinforced this idea. "This tumor is invading the brain," one surgeon murmured to another in an operating room. (By contrast, who ever said that the cold catches you?) Subject, verb, object: cancer was the autonomous actor, the aggressor, the mover. The hosts—the patients, their organs—were the hushed audience, the afflicted victims, the passive onlookers.

This language reflected an almost ontological commitment. It persisted even when research paradigms shifted. "Cancer is a genetic disease at its core," the MIT cancer biologist Robert Weinberg says. For decades, accordingly, biologists have looked for gene mutations that enable some aspect of cancer cells' aberrant growth, metabolism, regeneration, or behavior. In the late 1980s, a number of cancer biologists, Weinberg most prominently among them, threw themselves into finding such genes for metastasis—met genes, in effect. Might a breast-cancer cell, say, acquire a mutation that allowed it to unmoor itself from the breast and colonize the brain?

Despite a decades-long search, the met genes never materialized. "We looked and looked again, but we never found any,"

Weinberg told me. Occasionally, mutations were detected in cancer metastases that were different from the primary tumor, but no mutations emerged as singular drivers of metastasis. Starting in the late 1990s, cancer geneticists tried another approach. Mutations in cancer cells don't act in isolation; they can turn dozens, even hundreds, of other genes on and off. And those patterns of activation and repression can make an enormous difference—in the way that similar keyboards can produce wildly different sounds. (A caterpillar has the same genome as the butterfly it turns into, just as your liver cells have the same genome as your brain cells.) Instead of hunting for individual mutations, researchers looked for patterns of gene regulation—so-called gene-expression signatures. These patterns were used to develop predictive tests, which were rapidly shepherded into clinical trials.

For some variants of breast cancer, the tests turned out to be useful. Widely used gene-expression assays, such as MammaPrint and Oncotype DX, have helped doctors identify certain patients who are at low risk for metastatic spread and can safely skip chemotherapy. "We've been able to reduce the overuse of chemotherapy in about one-third of all patients in some subtypes of breast cancers," Daniel Hayes said.

Hayes is also grateful for the kind of genetic tests that indicate which patients might benefit from a targeted therapy like Herceptin (those whose breast cancers produce high levels of the growth-factor receptor protein HER2) or from antiestrogen medications (those whose tumors have estrogen receptors). But, despite our advances in targeting tumor cells using genetic markers as guides, our efforts to predict whose cancers will become metastatic have advanced only slowly. The "whether me" question haunts the whole field. What the oncologist Harold Burstein calls "the uncertainty box" of chemotherapy has remained stubbornly closed.

In 2001, Joan Massagué, a cancer biologist at New York's Memorial Sloan Kettering Cancer Center, came upon a scientific paper that radically changed his thinking about metastasis. Originally from Barcelona, Massagué—with his salt-and-pepper hair, his customary button-down shirt with an open collar—resembles a diplomat after embassy hours. He had spent years studying cell biology, elucidating mechanisms of gene regulation that might prime breast cells to travel to the bone instead of to the brain. Then came a crucial piece of evidence, buried in an obscure journal and published nearly three decades earlier. Researchers at the National Institutes of Health had implanted a sac of breast-cancer cells into the ovarian pedicle of a female rat. The cells grew to form a bean-sized tumor. The researchers then cannulated a large vein that was draining the tumor and siphoned blood from the vein every few hours in order to count the number of cancer cells that the tumor was shedding.

The results baffled the investigators. On average, they found, the tumor was sloughing off 20,000 cancer cells into every milliliter of blood—roughly 3 million cells per gram of tumor every 24 hours. In the course of a day, the tumor molted nearly a tenth of its weight. Later studies, performed with more sophisticated methods and with animal tumors that had arisen more "naturally," confirmed that tumors continually shed cells into circulation. (The rate of shedding from localized human tumors is harder to study, but available research tends to confirm the general phenomenon.)

"We imagine metastasis as a *going* problem," Massagué told me. "Mets go to the bone. Mets go to the brain." He punctuated the air with his fingers at each verb, his face flushed with excitement. "And—yes, yes—*going* is important, because we need to find what allows cells to break away from the tumor and enter the blood and the lymph nodes. But if primary human tumors shed cells continually, and if every cell is capable of forming visible metastasis, then every patient should have countless visible metastatic deposits all over his or her body." Anna Guzello's breast tumor should have stippled her brain, bones, and liver with mets. Why, then, did she have no visible evidence of disease anywhere else in her body? The real conundrum wasn't why metastases occur in some cancer patients but why metastases don't occur in all of them.

"The only way I could explain the scarcity of metastasis," Massagué said, "was to imagine that an enormous wave of cellular death or cellular dormancy must restrict metastasis. Either the cells shed by the tumor are killed, or they stop dividing, becoming dormant. When tumor cells enter the circulation, they must perish almost immediately, and in vast numbers. Only a few reach their destination organ, such as the brain or the bone." Once they do, they face the additional problem of surviving in unfamiliar and possibly hostile terrain. Massagué inferred that those few survivors must lie in a state of dormancy. "A visible, clinical metastasis—the kind that we can detect with CAT scans or MRIs—must only occur once a dormant cell has been reactivated and begins to divide," he said. Malignancy wasn't simply about cells spreading; it was also about *staying*—and flourishing—once they had done so.

In the spring of 2012, while Massagué and others were searching for sleeper cells, Gilbert Welch, an epidemiologist at Dartmouth, was preoccupied with a different problem: the unfulfilled promise of early detection. Early detection programs aimed to catch and eliminate cancers that were otherwise destined to become metastatic, but a huge ramp-up in screenings for certain cancers hadn't yielded comparable benefits in the mortality statistics. Welch was trained as a statistician as well as a physician, and when he recites numbers and equations his voice rises to a booming pitch, as if he were a televangelist moonlighting as a math teacher. To illustrate an extreme version of the problem, Welch told me the story of an epidemic-that-wasn't. In South Korea, starting about 15 years ago, doctors began to screen aggressively for thyroid cancer. Primary-care offices in Seoul were outfitted with small ultrasound devices, and doctors retrained themselves to catch the earliest signs of the disease. When a suspicious-looking nodule was found, it was biopsied. If the pathology report was positive, the patient's thyroid gland was surgically removed.

The official incidence of thyroid cancer—in particular, a subtype termed papillary thyroid cancer—began to soar across the nation. By 2014, thyroid-cancer incidence was 15 times what it was in 1993, making it the most commonly diagnosed cancer in the country. It was as if a "tsunami of thyroid cancer," in the words of one researcher, had suddenly hit. Billions of Korean wons were poured into treatment; tens of thousands of resected thyroids ended up in surgical buckets. Yet the rate at which people died from thyroid cancer remained unchanged.

What happened? It wasn't medical error: observed under the microscope, the questionable nodules met the criteria for thyroid cancer. Rather, what the pathologists were finding wasn't particularly pathological—these thyroid cancers had little propensity to cause illness. The patients had been not misdiagnosed but overdiagnosed; that is, cancers were identified that would never have produced clinical symptoms.

In 1985, pathologists in Finland assembled a group of 101 men and women who had died of unrelated causes—car accidents or heart attacks, say—and performed autopsies to determine how many harbored papillary thyroid cancer. They cut the thyroid glands into razor-thin sections, as if carving a hock of ham into prosciutto slices, and peered at the sections under a microscope. Astonishingly, they found thyroid cancer in more than a third of the glands inspected. A similar study regarding breast cancer— comparing breast cancer incidentally detectable at autopsy with the lifetime risk of dying of breast cancer—suggests that a hyper-zealous early detection program might overdiagnose breast cancer with startling frequency, leading to needless interventions. Survey-ing the results of prostate-cancer screening, Welch calculated that 30 to 100 men would have to undergo unnecessary treatment— typically, surgery or radiation—for every life saved.

"The early detection of breast cancer via mammography saves women's lives, although the benefit is modest," Daniel Hayes told me. But equally important is the question of what to do with the tumor we've detected: can we learn how to identify those cancers that need to be treated systemically with chemotherapy or other in-terventions? "It's not just early detection that we want to achieve," Hayes went on. "It's early *prediction*."

For Welch, the fact that diagnoses of thyroid cancer or pros-tate cancer could soar without a corresponding effect on mortal-ity rates was a warning: a little knowledge had turned out to be a dangerous thing. Cancer-screening campaigns had expanded the known reservoir of disease without telling us if, in any particular case, treatment was necessary. Early detection helped us with *when* and *what* but not with *whether*. And there was an element of mys-tery. Why did some cancers spread and kill patients, while many remained docile?

One day in March 2012, Welch flew to Washington to attend a conference on cancer metastasis. It was a gusty, gray morning— "the hotel was nondescript, the food unremarkable"—and Welch, dangling the requisite nametag on a forlorn lanyard, found him-self in a room full of cancer biologists, feeling like an alien species. "I study patterns and trends in cancer in human populations," he told me. "I take the 100,000-foot view of cancer. This meeting was full of metastasis biologists looking at cancer cells under the mi-croscope. I couldn't tell what any of this had to do with population trends in human cancer—or, for that matter, why I'd even come to this meeting."

Then, coffee jolting in his hand, he saw a slide on the screen that made him sit up and take notice. It depicted the infestation of mussels in Lake Michigan. The speaker, Kenneth Pienta, an oncologist from the University of Michigan (and now at Johns Hopkins), had heard about the quagga crisis and been struck by the seeming parallels with cancer. Rather than viewing invasiveness as a quality intrinsic to a cancer, researchers needed to consider invasiveness as a pathological relationship between an organism and an environment. "Together, cancer cells and host cells form an ecosystem," Pienta reminded the audience. "Initially, the cancer cells are an invasive species to a new niche or environment. Eventually, the cancer-cell–host-cell interactions create a new environment." Ask not just what the cancer is doing to you, Pienta was saying. Ask what you are doing to the cancer.

By talking about cancer in ecological terms, Pienta was, in the tradition of Paget and Fidler, urging his colleagues to pay more attention to the soil. A woman with a primary tumor in her breast was caught in a pitched but silent battle. Oncologists had spent generations studying one possible outcome of that battle: when the woman lost, she succumbed to metastasis. But what happened when cancer lost the battle? Perhaps cancer cells tried to invade new niches but mainly perished en route, as a result of the resistance mounted by her immune system and other physiological challenges; perhaps the select few that, singly or in clusters, survived the expedition ended up languishing in forbidding tissue terrain, like seeds landing on a salt flat.

Welch was captivated. We had to be alert to the differences between the rampaging quagga mussel and the endangered purple-cat's-paw mussel—but what about the differences between the Great Lakes and the Dnieper? Evidence suggested, for example, that most men with prostate cancer would never experience metastasis. What made others susceptible? The usual approach, Welch knew, would be to look for markers in their cancer cells—to find patterns of gene activation, say, that made some of them dangerous. And the characteristics of those cells were plainly crucial. Pienta was arguing, though, that this approach was far too narrow. At least part of the answer might lie in the ecological relationship between a cancer and its host—between seed and soil.

*

In 1992, an Australian high school teacher in his late fifties was diagnosed with melanoma. The malignancy began as a streak of black—a cancellation sign extending from his left armpit across the torso. A few weeks after the diagnosis, though, the borders of the tumor began to change. One edge turned gray; another shrank. "He had a classic spontaneous regression—typically a sign that the cancerous lesion was being controlled by the immune system," David Adams, the man's son, told me. The primary melanoma was surgically resected, and no metastasis was ever found. One of his father's friends, also in his fifties, was not so lucky: by the time his primary melanoma had been discovered, his brain was sprinkled with visible mets.

David Adams went on to train as a geneticist and a physiologist in Sydney, before joining the Sanger Institute, in Cambridge, England. There he leads a group studying the biology of melanoma. Originally from Tamworth, a small outback town in New South Wales ("hot, flat farming country, right in the middle of Australia's melanoma belt," he says), Adams now lives 10,000 miles away, in a quaint English village, speaks with a mild Cantabrigian accent, and drives a gently distressed compact car to work. He has, in short, gone native—a matter of soil over seed, you might think —but he hasn't forgotten his father's case; it's what has driven his scientific career. What had made a melanoma regress in one host and turn aggressive in another? Adams knew of a strange series of melanoma cases, occasionally reported in the medical literature, involving donated kidneys. They fit a pattern. A patient—call him D.G.—is diagnosed with a melanoma and successfully treated with surgical resection. Years later, D.G., now deemed perfectly healthy, donates a kidney to a friend. The friend is prescribed routine immune suppressants to prevent the rejection of the kidney. A few weeks later, however, the recipient begins to sprout hundreds of black pinpricks of melanoma in the kidney. The melanoma, bizarrely, has come from D.G.'s cells. The donated kidney has to be removed. Meanwhile, the donor—like some Dorian Gray of transplantation—remains uncannily healthy, with no sign of melanoma in his body.

Here, too, Adams realized, the original host environment played a crucial role in restricting metastatic growth. The donor's melanoma cells must have been sitting dormant in the donated kidney, akin to the phenomenon of dormancy that Massagué had found in mice. When the "soil" changed, and the dormant cells arrived in an immune-suppressed recipient, the cancer began to grow. "The immune response in the donor must have been restricting the metastatic cancer's growth," Adams told me.

In 2013, Adams began to conceive an ambitious experiment to identify cancer-suppressing host factors. "Just a few yards from my office, there is an animal vivarium filled with hundreds of genetically altered mouse strains," he said. "Researchers were using these strains to study the effect of these gene variants on the heart, or on the nervous system. I thought I would ask a somewhat different question: if we implanted these strains with the same cancer, which strains would permit the metastases to grow, and which ones would suppress metastatic outgrowth?"

It was an ingenious inversion of a classic experimental strategy. For decades, biologists have been altering a cancer cell's genes and injecting the cells into a few standardized strains of mice. The "different cancers into same strain" experiments have allowed cancer biologists to observe how alterations in cancer genes might affect their growth, metabolism, and metastasis. But what effects might variations in the host's genome have? Adams's "same cancer into different strains" experiment switched the locus of attention from seed to soil.

In New York and Boston, meanwhile, researchers such as Joan Massagué and Robert Weinberg were also investigating "host factors." In a suggestive experiment, Weinberg and his colleagues studied a cohort of mice whose lungs they had sprayed with thousands of dormant cancer cells. Some mice were exposed to an inflammatory stimulus—the kind that might occur during pneumonia, say—and only in those did the "micro-mets" wake up and turn aggressive. It called to mind a fascinating, if overlooked, experiment that Mina Bissell had done back in the 1980s. Researchers had known for generations that if you injected a chick's wing with a certain cancer-causing virus a tumor would grow there. Bissell showed that, when you injected one wing and injured the other, this other wing would grow a tumor, too. On the other hand, if you injected a chick while it was an embryo, there would be no tumor at all. "Back then, it was fashionable to think of cancer only as an oncogene-driven automaton," Bissell told me. "But here the

automaton could be switched on and off by its local environment."
It wasn't just the seed that mattered; changing features of the soil
could affect whether it would ever germinate.

Massagué and his students were making advances of their own,
notably in an experiment in which they depleted various types of
immune cells in mice that carried dormant cancer cells. Some of
these cell types belong to the "adaptive immune" system, which
learns to identify new pathogens and to target them when they
next appear. (The adaptive immune system, associated with T
cells and B cells, is why vaccines work, and why people seldom
get chicken pox more than once.) But the most striking effect
occurred when the experimenters depleted another type of cell,
the "natural killer," or NK, cell. These cells belong to our "innate
immunity"—they can't learn anything new but arrive prepro-
grammed to destroy sick or aberrant host cells. Massagué's team
had implicated these cells as crucial surveyors and controllers of
cancer metastasis.

Adams's particular interest was in host genes, rather than cell
types, that might affect metastasis. In early 2013, Louise van der
Weyden, a postdoc in Adams's lab who also happens to be his
wife, created a suspension of mouse melanoma cells—a coffee-
dark slurry—and injected it into a few dozen mouse strains. Some
weeks later, she counted the number of visible mets in the lungs
for each strain and rushed the data to Adams's office.

Even within that small cohort, Adams recalled, the differences
were obvious. Some of the mice had developed hundreds of mets
—a fusillade of black pinpricks. In still others, the lungs had visibly
blackened with metastasis. Yet some mice had developed just a few
mets. Adams has a photograph of those mouse lungs above his
desk. "Here was the same cancer exerting such different effects in
different host environments," he said.

Two years later, van der Weyden had inoculated 810 mouse
strains with the melanoma cells and scrutinized the physiology of
metastasis in each. Fifteen strains were either moderately or extremely resistant. Twelve of those 15 strains had gene variations
that affected immune regulation, again suggesting the potent
role of that system in a cancer's ability to spread and invade. Even
within the resistant group, one mouse strain stood out. Exposed to
the dose of cancer cells used in the study, normal mice developed
about 250 mets. Mice of this resistant strain, however, developed

only 15 to 20 mets on average. And some of these mice hardly developed any mets at all; their lungs looked pristine and uncolonized even two months after the exposure.

Was this resistance to metastasis peculiar to melanoma, which is a type of cancer well known to provoke an immune response? Adams and van der Weyden tested three other types of cancer: lung, breast, and colon. In all of them, the mouse strain was resistant to the formation of metastases. Notably, the strain carries a variant in a gene called Spns2, which, through a cascade of events, increases the concentration of immune cells, notably NK cells, in the lungs —the very cells that Massagué's lab had identified as a powerful restrictor of metastasis.

David Adams's father never suffered a recurrence of melanoma; he died from prostate cancer that had spread widely through his body. "Years ago, I would have thought of the melanoma versus the prostate cancer in terms of differences in the inherent metastatic potential of those two cell types," Adams said. "Good cancer versus bad cancer. Now I think more and more of a different question: why was my father's body more receptive to prostate metastasis versus melanoma metastasis?"

There are important consequences of taking soil as well as seed into account. Among the most successful recent innovations in cancer therapeutics is immunotherapy, in which a patient's own immune system is activated to target cancer cells. Years ago, the pioneer immunologist Jim Allison and his colleagues discovered that cancer cells used special proteins to trigger the brakes in the host's immune cells, leading to unchecked growth. (To use more appropriate evolutionary language: clones of cancer cells that are capable of blocking host immune attacks are naturally selected and grow.) When drugs stopped certain cancers from exploiting these braking proteins, Allison and his colleagues showed, immune cells would start to attack them.

Such therapies are best thought of as *soil* therapies: rather than killing tumor cells directly, or targeting mutant gene products within tumor cells, they work on the phalanxes of immunological predators that survey tissue environments, and alter the ecology of the host. But soil therapies will go beyond immune factors; a wide variety of environmental features have to be taken into account. The extracellular matrix with which the cancer interacts, the blood vessels that a successful tumor must coax out to feed itself, the nature of a host's connective-tissue cells—all of these affect the ecology of tissues and thereby the growth of cancers.

Cancers, like mussels, proliferate in congenial habitats, and, like mussels, they can create microenvironments that help them resist predators. Seed therapies kill cells—something like spraying a lake with a mussel poison. Soil therapies, by contrast, change the habitat. When I asked Adams about the kind of clinical trial that excited him because of its therapeutic potential, he discussed an unusual study in which patients who are diagnosed with a primary melanoma—such as his father—will donate blood so that researchers can identify their genetic markers and their immune-cell composition. By studying how they fare over time, we might learn which patient populations are particularly susceptible or resistant to certain cancers. We'd have a better sense of which patients need aggressive treatment. And we might learn something about *how* to treat them—how to alter a susceptible patient's immunological and histological profile to resemble that of a resistant one.

"Cancer is no more a disease of cells than a traffic jam is a disease of cars," the British physician and cancer researcher D. W. Smithers wrote in *The Lancet*, in 1962. "A traffic jam is due to a failure of the normal relationship between driven cars and their environment and can occur whether they themselves are running normally or not." Smithers had overstepped in his provocation. The uproar that ensued was clamorous and immediate; Smithers complained that he had been "lacerated by Occam's razor." By arguing that cellular *relationships* were responsible for cancer's behavior, he had committed the cardinal sin of multiplying the factors that oncologists had to consider. "To deny the importance of cells in tumor growth would be like denying the importance of people in some problem in sociology," he later clarified. Cancer cells were a necessary condition for disease but not a sufficient one. His real aim was to get beyond oncology's obsession with its internal-combustion engine—the cellular automaton and its genes—and only since his death has the field started to come to grips with his message.

You ride the subway one morning. The train is delayed at 59th Street, and a man in a Yankees cap sneezes on you. At work later that week, you feel the chill entering you quietly, on little cat feet.

You take a cab home, now sniffling, cursing the C line and retracing your steps: the culprit with the cap; the empty seat that should have raised suspicion; that slightly moist steel bar you should never have touched. What you do not think about are the six other passengers, sitting nearby, who also got sneezed on. None of them is sick.

This is medicine's "denominator problem." The numerator is you—the person who gets ill. The denominator is everyone at risk, including all the other passengers who were exposed. Numerators are easy to study. Denominators are hard. Numerators come to the doctor's office, congested and miserable. They get blood tests and prescriptions. Denominators go home from the subway station, heat up dinner, and watch *The Strain*. The numerator persists. The denominator vanishes.

Why didn't the denominators get sick? The pathogen exposure was the same; the hosts were different. Yet even the term *pathogen* is misleading. A pathogen is defined by its ability to be, well, pathogenic. That's not an inherent attribute, however; it's a relationship, an interaction with the host. Ruslan Medzhitov, an immunobiologist at Yale, has spent much of his life studying host-pathogen interactions. "You can inject the same virus into different hosts and get vastly different responses," he says. It's the soil that determines the nature of the illness.

And that returns us to the problem with the early detection paradigm. Suppose we could install tiny sensors in people that would regularly scan their blood to find circulating tumor cells, conducting an ongoing "liquid biopsy." We'd be catching cancers earlier than ever before. But, as with the doctors in Seoul, we might also end up overtreating more cancers than ever before. That's because circulating tumor cells might augur metastatic cancer in some patients, while in others the mets never seem to take hold. Why don't the mets take hold? The old answer was: the cancer wasn't the right kind of guest. The new question is: should we be looking, too, for the right kind of host?

A few months ago, a forty-year-old woman came to my office in a state of panic. She had had a hysterectomy as a treatment for endometriosis. Pathologists, examining her uterus postoperatively, had found a rare, malignant sarcoma lodged in the tissue—a tumor so small that it could not be seen on any of her preoperative scans. She had consulted a gynecologist and a surgeon, both of whom had recommended an aggressive procedure to remove the ovaries and the surrounding tissue—a scorched-earth operation with many long-term consequences. Once these tumors spread, they had reasoned, there's no known treatment. Patients diagnosed with these sarcomas tend to have a sobering prognosis, with most surviving only two to three years after the symptoms appear.

But that's a completely different scenario, I said to her. In her case, the tumor was detected incidentally. There were no symptoms or signs of the cancer. If we sampled 10,000 asymptomatic women, we have no idea how many such malignancies would be found incidentally. And we have no clue how those tumors, the ones found incidentally, behave in real life. Would the alliances formed between the woman's tumor cells and her tissue cells enable widespread metastatic dissemination? Or would these encounters naturally dampen the growth of the tumor and prevent its spread? Nobody could say. We err toward risk aversion, even at the cost of bodily damage; we don't learn what would happen if we did nothing. It was a classic "denominator" problem, but my response seemed supremely unsatisfactory.

She looked at me as if I were mad. "Would *you* sit and do nothing if someone found this tumor in you?" she asked. She decided to go ahead with the surgery.

Anna Guzello went in the opposite direction, as I recently learned when I checked back with her oncologist, Katherine Crew. Guzello had agreed to take the estrogen-blocker tamoxifen. But she refused chemo, and even Herceptin, despite being HER2-positive. Frustratingly, though, Crew wasn't in a position to say with any confidence what was going to happen.

For decades, our standard explanation for those who make up our "denominators"—i.e., people who meet the criteria of the diagnostic test, who are at risk for a disease, but who may not actually have it—was stochastic: we thought there was a roll-of-the-dice aspect to falling ill. There absolutely is. But what Medzhitov calls "new rules of tissue engagement" may help us understand why so many people who are exposed to a disease don't end up getting it. Medzhitov believes that all our tissues have "established rules by which cells form engagements and alliances with other cells." Physiology is the product of these relationships. So consider our internal-denominator problem. There are tens of trillions of cells in a human body; a large fraction of them are dividing, almost always imperfectly. There's no reason to think there's a supply-side shortage of potential cancer cells, even in perfectly healthy people. Medzhitov's point is that cancer cells produce cancer—they get established and grow—only when they manage to form alliances with normal cells. And there are two sides (at least) to any such relationship.

Once we think of diseases in terms of ecosystems, then, we're obliged to ask why someone *didn't* get sick. Yet ecologists are a frustrating lot, at least if you're a doctor. Part of the seduction of cancer genetics is that it purports to explain the unity and the diversity of cancer in one swoop. For ecologists, by contrast, everything is a relationship among a complex assemblage of factors.

I talked to Anthony Ricciardi, professor of invasion ecology at McGill University, in Montreal. Ricciardi, a biologist, grew up on the banks of Lake Saint-Louis, which bulges out from the Saint Lawrence River—the route through which the mussels metastasized to the Great Lakes. "I was familiar with much of what was living in that lake, having played in it as a child and later studied it as a student," he told me. "And I had never seen a zebra mussel before. Then, one day in June 1991, while I was working on a research project, I turned over a rock and there was one of them attached to it. It took me a few seconds to recognize what it was. And then I found a few more. That's when I had a premonition of the invasion to come."

I asked him why those freshwater mussels went into hyperdrive when they came to our lakes. "You've got to understand the dynamics of invasion ecology," he said. "It's a series of dice rolls. Most organisms introduced into a new environment will fail, often because they arrive in the wrong place at the wrong time. Vast, vast numbers will die. Piranhas were dumped into the lake for years, but they can't establish, because the temperature isn't right for them. People will release marine species like flounder, but the salinity isn't right for them." His language, even his tone, was eerily reminiscent of Joan Massagué's; he might have been describing the waves of cellular death during the establishment of metastasis. "There isn't one factor but a series of factors that determined how and why the mussels took hold," he went on.

"But, over all, would you say the temperature of the water was the key?" I asked.

"The water temperature's a factor. The water chemistry would also have contributed."

"So a combination of the temperature and the salinity?"

"But also the calcium content. That's absolutely important."

I added that to my list of drivers: "Temperature, salinity, calcium . . ."

"And the fact that there weren't any well-adapted predators. The native fish in these lakes will hardly touch the mussels. Neither will most ducks."

"Ducks?"

He sighed, as if tasked with explaining an immensely complex theorem to a child. "There are many contributing factors, although some of these factors are clearly more important than others. There are probabilities attached. It's all context-dependent."

And so it went. For a cancer geneticist like me, it was an exercise in frustration. Every time I tried to pin down a principal cause for the *Dreissena* invasion, I was presented with another contender. Disheartened, I gave up.

Perhaps we all gave up. Considering the limitations of our knowledge, methods, and resources, our field may have had no choice but to submit to the lacerations of Occam's razor, at least for a while. It was only natural that many cancer biologists, confronting the sheer complexity of the whole organism, trained their attention exclusively on our "pathogen": the cancer cell. Investigating metastasis seems more straightforward than investigating non-metastasis; clinically speaking, it's tough to study those who haven't fallen ill. And we physicians have been drawn to the toggle-switch model of disease and health: the biopsy was positive; the blood test was negative; the scans find "no evidence of disease." Good germs, bad germs. Ecologists, meanwhile, talk about webs of nutrition, predation, climate, topography, all subject to complex feedback loops, all context-dependent. To them, invasion is an equation, even a set of simultaneous equations.

Still, at the ASCO meeting this June, on the shore of Lake Michigan, I was struck by the fact that seed-only research was increasingly making room for research that also sifted through soil, even beyond the excitement surrounding immune therapies. Going further and embracing an ecological model would cost us clarity. But over time it might gain us genuine comprehension.

Taking the denominator problem seriously beckons us toward

a denominator solution. In the field of oncology, "holistic" has become a patchouli-scented catchall for untested folk remedies: raspberry-leaf tea and juice cleanses. Still, as ambitious cancer researchers study soil as well as seed, one sees the beginnings of a new approach. It would return us to the true meaning of "holistic": to take the body, the organism, its anatomy, its physiology—this infuriatingly intricate web—as a whole. Such an approach would help us understand the phenomenon in all its vexing diversity; it would help us understand when you have cancer and when cancer has you. It would encourage doctors to ask not just what you *have* but what you are.

KIM TODD

The Island Wolves

FROM *Orion*

IN THE BONE GARDEN, moose antlers sprout like pale stalks of mutant corn. Skulls, lined up one behind the other, bear tags that sketch moose lives: "2 yrs. old starved, winter '96" and "Died 1989. Wolves dug through 2 ft. of snow to feed on it in 1990." The curious visitor to this small plot on Lake Superior's Moskey Basin in Isle Royale National Park can also marvel at thematic displays: worn skulls by a sign declaring "The Odd Fellows corner—old bulls no longer reproducing, antlers in decline" and shelves of the skulls and antlers of dead young males labeled "Darwin Award Winners."

No wolf bones lie here—they have been sent to a lab at Michigan Technological University with thousands of moose metatarsals —but the Bone Garden's skulls and antlers are a library of wolf sagas written in gnawed jawbone. Isle Royale, an isolated wilderness of boreal forest and sheltered swamp surrounded by a rocky shoreline and dotted with plentiful loons, is the site of the longest-running study of predator-prey dynamics in history, spanning almost six decades. To ecologists, these bones tell some of the most famous animal stories in the world, framing ideas about what it means to eat and be eaten, and how the battle to survive shapes the land.

But now, it appears the stories have been misinterpreted.

"I was dead wrong," says Rolf Peterson, who has been studying wolves and moose on Isle Royale since 1970. Peterson, dressed in a plaid shirt and down jacket patched with duct tape, gray beard still streaked with a hint of red, gives me the garden tour, pointing

out two skulls locked together in what must have been a fight to
the death. His speech is slow and thoughtful, marked by a rueful
chuckle.

For years, Isle Royale National Park has been considered the
perfect natural lab: a closed system with one dominant predator,
the wolf, relying heavily on one kind of prey, the moose. A 45-mile-
long island in one of the world's largest lakes, Isle Royale is defined by solitude. Miles from the mainland, in the summer, the
park can only be reached by seaplane, private boat, or a half-day
ferry ride from either Michigan or Minnesota; in winter, the park
is closed, abandoned to deep snows and wildlife. Though few in
numbers, the wolves marooned in the park thrived for years, seemingly free of the deformities and stillbirths that plague populations
where close relatives breed. Recording wolf and moose numbers
and documenting hunting and fleeing, biologists charted two species intimately intertwined. But recent revelations resulting from
DNA tests took everyone by surprise.

"You think if you watch a place for a long time, you'll have a
deeper understanding," says John Vucetich, lead scientist of the
wolf-moose study. But the new information upended expectations.
"It's not as though it was an accumulation of knowledge; it was a
full-scale reversal of what we know."

For centuries on Isle Royale, there were no wolves, just caribou
lurking in the balsam fir and lynx chasing snowshoe hares over
the stone outcroppings, until hunters and trappers cleared them
out. Moose arrived on the island by the early 1900s, probably by
swimming, and lived there for about 50 years without any predators. Then the wolf rumors started, spurred by possible signs: an
outsized paw print on a beaver dam. Scat laced with moose hair.

In the early 1950s, Lee Smits, a newspaperman and conserva-
tionist, saw Isle Royale as the perfect spot for a wolf refuge, a place
the predator could be safe from the poison, guns, and traps that
killed off most of its kind in the Lower 48. There were no ranchers
to anger or sheep to harass. Maybe wolves would find their way to
the island themselves. Maybe they already had, Smits thought. But
he wanted to be sure. He asked trappers on Michigan's mainland
to keep an eye out for pups that could be introduced to Isle Royale
to start a pack. When they didn't find any, he turned to the Detroit
Zoo.

Lady, Queenie, Adolph—the chosen wolves had names. The fourth was Big Jim, a bulky male descended from Saskatchewan stock and bottle raised by Smits and his wife. Big Jim learned to retrieve ducks by playing with their water spaniel. Not surprisingly, when brought to the island in August 1952, the zoo wolves didn't know how to behave. Instead of chasing down moose, they chewed up fishing nets, raided laundry lines, and hung out near the Rock Harbor Lodge. As a *New York Times* headline summed it up, "The Wild Calls in Vain to Four Urbane Wolves."

Rangers hauled them to a more desolate spot, but they came back to the lodge, startling tourists. By then, the rangers were fed up. They shot Adolph. And then Queenie. They captured Lady and took her back to the zoo. Big Jim escaped. He turned up here and there over the years. In 1957, a park service biologist reported a wolf trailing him like a wary dog. But no one much cared, because before long there was solid evidence of other wolves, too, wild ones. A year later, Durward Allen, a professor at Purdue University, and his graduate student L. David Mech launched a study of these wolves, the moose, and their entanglement.

The scientists' conclusions quickly found their way into textbooks about predators and prey. Pick one up, even today, and you will likely find pictures from Isle Royale: wolves like black bullets tearing through snow, biting at the heels of a calf stumbling in the deep drifts. The spray of blood and exposed spine on a creek bank. A whole pack, photographed from above, spreading out over the frozen lake, scribbles of tracks and shadow, like graphs on white paper.

In these textbooks you will also find actual graphs—all this behavior abstracted, slobbering tongues and flicking whiskers and clumps of hair smoothed away—because they illustrate the lesson so neatly. Allen and Mech's research fit perfectly into the Lotka-Volterra predator-prey model. Developed in the 1920s, the model is a set of differential equations that show the predictable way the fates of species are tied together. As the prey population grows, the predators, finding more food, also flourish. As young are eaten and adults harried, the prey population falls. The predator population then also falls, as their food grows scarce. The equations produce graceful waves, one trailing the other—a field mouse leaping, a fox leaping after.

In his book *Sociobiology*, Edward O. Wilson described Isle Royale and these early studies as "a simple and instructive example of the balance between predator and prey." He highlighted the benefits of this balance for the prey, often pitied as victims:

> The moose, by unwillingly supplying the wolves with one of their members about every three days, have stabilized their own population . . . As a curious side effect, the moose herd is kept in good physical condition, since the wolves catch mostly the very young, the old, and the sickly individuals. And, finally, because the moose population is not permitted to increase to excessive levels, the vegetation on which they feed remains in healthy condition.

In Wilson's description, the orderly cycles gestured at an irresistible, uplifting tale. The Isle Royale study has the same appeal as a Jack London novel—harsh beauty, high drama, and, importantly, an explanation for the violence of the natural world.

As a high school student in Minneapolis, Rolf Peterson read about Durward Allen's work in the *Star Tribune* and plastered newspaper photos of Isle Royale foxes on his wall. A 1963 *National Geographic* article, "Wolves Versus Moose on Isle Royale," further whet his appetite. It showed a golden-eyed wolf staring down the research plane used for winter counts and a pack swarming at a faltering cow moose. It described wolves howling, a sound few in the continental United States had ever heard. As a college senior in 1969, Peterson wrote to Allen at Purdue and asked to be part of the study. Allen said yes.

Peterson tells me this story at the kitchen table of the weathered Bangsund log cabin that, for decades, has served as the base of operations for the Isle Royale study. The cabin has bright red trim and a sign on the door that reads, THIS IS OLD GLASS. PLEASE DO NOT LET THE DOOR SLAM IN THE WIND. Inside is a small wood stove. Island maps cover the wall, the table, and the low, sloping ceiling. The decor consists of a howling wolf in stained glass, a hanging moose skull, and beaver skulls on the windowsill. The overall aesthetic is that of a field-research station crossed with *Little House in the Big Woods*. At night the cabin glows with the light of solar-powered paper lanterns.

Peterson traces wolf-pack movements on the map on the table, takes out the 1963 *National Geographic,* and reads lines near the end of the article about the island: "Our studies thus far indicate that the moose and wolf populations on Isle Royale have struck a reasonably good balance." The wolf numbers had remained stable, at around 20 or so, and the moose numbers, as best as they could count them, seemed to be holding steady too. "The notion of balance was firmly ingrained in everything," Peterson says. When Peterson joined the study, it had already been running for 12 years, and the chief ranger said to him, "You're in the 12th year of a 10-year study. What are you going to do?"

But Allen told Peterson, "Just watch, something will happen." And it did.

During Peterson's first summer season, everything began to break loose. After a series of harsh winters, wolf numbers soared, going up to 50 in 5 packs at one point. Behavior changed, with wolves chasing moose off cliffs and eating just the organ meat before killing again. Moose numbers plummeted, then wolf numbers plummeted. Then wolf numbers rose again. What was causing such dramatic shifts?

As the scientists plotted the data, tidy Lotka-Volterra waves vanished, replaced by something else. "It's like a big historical novel. Every 5-year period looks completely unlike any of the previous 5-year periods, and the dynamics are driven by external events that we cannot imagine, let alone predict," Peterson says.

Though the biologists flew over in helicopters, recording all they could see, unseen forces governed the island. In 1997, one of the coldest winters of the century, the lake froze. Blizzards battered the Lake Superior shore, and a hulking wolf, who would grow silver as he aged, struck out from Ontario in the direction of Isle Royale.

He found an island reeling. In 1981, an infected dog from Chicago, brought in illegally on a private boat, had carried canine parvovirus to the island. No pups survived that year, and the wolves had not entirely bounced back. Two years before the lone wolf's arrival, the moose, up to almost 2,500 head, had chewed balsam fir to nubs and starved, crashing down to about 500 individuals. Wolves suffered as a result: old ones died, pups died, and those who scraped through did so on empty stomachs. Hungry survivors gnawed the hide of a moose that had been dead for a full year.

Unconcerned by the bleak conditions, the newcomer took over. With him at the head, the Middle Pack, one of three packs on the island at the time, aggressively expanded beyond its moose-poor territory, claiming richer areas such as Chippewa Harbor. The new wolf raised six pups in 1998 and three the next year. The island's wolf population surged. Noting a clear new pack leader with increasingly pale fur, and not knowing he came from the mainland —not knowing any wolves still came over from the mainland—the biologists dubbed him Old Gray Guy.

The idea of a balance in nature is at least as old as Herodotus, the Greek historian, who tried to understand why prey animals weren't entirely consumed. Investigating, he gathered reports from the Middle East: "Of a truth, Divine Providence does appear to be, as indeed one might expect beforehand, a wise contriver. For timid animals which are a prey to others are all made to produce young abundantly, that so the species may not be entirely eaten up and lost; while savage and noxious creatures are made very unfruitful." For many years, this sense of balance was the hinge between religion and biology. William Derham wrote in his 1713 book *Physico-theology: or, A Demonstration of the Being and Attributes of God, from His Works of Creation* that divine management can be seen in the "Adjustment of the Quantity of Food to the Number of Devourers, so that there is not too much, so as to rot, and annoy the World." Though underneath the cheerful accounting lay the niggling question: why was this violent, unending cycle part of the divine plan?

In the late nineteenth and early twentieth centuries, some writers came up with an answer: it wasn't. Plains were empty of bison, woods had only a few deer. Conservationists predicted that predators would do what Herodotus thought they couldn't: eat prey into extinction. Teddy Roosevelt, in his 1893 book *The Wilderness Hunter,* wrote, "The wolf is the arch type of ravin, the beast of waste and desolation." *Ravin* connotes robbery, greed, and plunder. William T. Hornaday, in *Our Vanishing Wildlife,* constantly referred to wolves as cruel, and said insatiable human hunters have "the gray-wolf quality of mercy"—by which he meant none. This language fed the vicious predator-extermination campaigns.

Against this wave of righteousness, author Ernest Thompson Seton built a seawall of romance. In his most famous story, "Lobo, the King of Currumpaw," the noble outlaw Lobo, who is the leader of a wolf pack, is done in by his love for the beautiful but ditzy

Blanca. The narrator, a wolf hunter who traps Lobo at last, finds himself moved to mercy: "Yet before the light had died from his fierce eyes, I cried, 'Stay, we will not kill him; let us take him alive to the camp.'" Lobo dies anyway.

Seton's books are the definition of charm. Ink sketches adorn the margins: a nest at the top of a tree that runs the whole length of the page, musical notes showing the various caws of the crow, a fox tugging a chain to release her son, a jotting of poison sumac, partridge tracks wandering over the paper field. The title type is like something burned into wood over a cabin door.

Seton's prose is as idiosyncratic as his books' design. In *Wild Animals I Have Known*, Seton lays out his methodology for what might be a book of natural history or a collection of short stories or a series of animal fables for children: "I believe that natural history has lost much by the vague general treatment that is so common. What satisfaction would be derived from a ten-page sketch of the habits and customs of Man? How much more profitable it would be to devote that space to the life of some one great man. This is the principle I have endeavored to apply to my animals."

Seton's weird notions about how to write about grizzlies and rabbits faded from popularity. But Lobo lived on in what is probably the most significant story of the American conservation movement, 1949's "Thinking Like a Mountain" by Aldo Leopold. The dying wolf at the end of Leopold's essay is an echo of Lobo: "We reached the old wolf in time to watch a fierce green fire dying in her eyes. I realized then, and have known ever since, that there was something new to me in those eyes—something known only to her and to the mountain." What she and the mountain know is that wolves keep the deer in check, deer that would otherwise destroy the mountain. The 1963 *National Geographic* Isle Royale article used Seton's character as a stand-in for all of his kind: "The truth is that Lobo is gone from nearly all his old haunts." For a long time, when someone said *wolf,* Lobo came to mind.

That is until the Isle Royale wolves replaced him in the American imagination, swapping out one wolf tale for another. Lobo is the story of an individual, heroic and tragically flawed. The Isle Royale wolves have long been interpreted as characters in a narrative about how a good system works. No less noble than Lobo in their way, they restore the balance of nature. In this update of the ancient idea, though, balance results from natural selection rather than a divine manager.

If Isle Royale is the textbook example of the predator-prey relationship, Yellowstone is the Imax movie. Only a few years after wolves were reintroduced there in 1995, they were credited with the park's transformation. They ate elk. This reduced elk browsing, allowing aspen and willow to shoot to heights they hadn't reached in decades. Healthy groves of aspen and willow, in turn, sheltered beavers and birds. This chain of effects is called a *trophic cascade,* where a predator sends ripples throughout the food web. It is a similar idea to that of a *keystone species*, where, for example, a sea star on the Washington coast, once removed, causes mussels to flourish and limpets to decline. According to this view, some species have an outsized effect on the entire landscape, shaping the ecosystem. But brush the soil off these notions and "balance of nature" appears as their bedrock: "Wolves keep Yellowstone in balance" declares a blogger on the website Canis Lupus 101.

Yet as Peterson and his colleagues counted wolves, counted moose, and tracked Old Gray Guy and his offspring through meadows and bogs, balance on Isle Royale proved increasingly elusive.

In February 2000, Rolf Peterson and his pilot saw a wolf cowering on a rock offshore as their plane skirted the coast. The Middle Pack waited on the bank to finish her off. Led by Old Gray Guy, the pack had become a powerhouse, taking over 75 percent of the island and killing a stray male from the East Pack two days before.

As Peterson watched, the wolves surged into the water, snarling and biting at the wolf on the rock, then retreated. After weathering a dozen attacks, the lone wolf jumped into the icy lake and swam to another rock, then another. The pack followed on land, rushing through the waves to tear at her whenever she paused. A photo from above shows the pack on a submerged stone, arrayed like a sea star around a point—the wolf in the center, downed and invisible.

Peterson, talking into his tape recorder, said three times, "I think she's dead now." But she kept getting back up. Finally, after making it to shore only to be attacked again, she seemed definitively gone.

The men eventually flew to Windigo for a body bag and to refuel. When they got back, a male who had been far behind his pack mates approached the carcass. And then, as Peterson later wrote in his report, "the lone female raised her head!" (See *The Ecological Studies of the Wolves of Isle Royale, 1999–2000.*) The next day, both wolves were gone, leaving behind bloody indents in the snow where the female had slept.

Several days later, Peterson found the pair. The female was standing up and the male was licking her neck. Every now and then he'd dig in the snow to imitate caching food—courting— and she'd growl, and he'd go back to licking her neck. The scientists dubbed her the Cinderella wolf, but Cinderella doesn't seem right. Maybe the second wolf staged a rescue, but invoking the fairy tale obscures the first wolf's own swimming and fighting and snarling determination to live. There was no fairy godmother or party dress. A better name for her is Ferocious Warrior. The male wolf—he can be Prince Charming.

At the end of his account in the 1999–2000 annual report, Peterson concludes with this description of the pair: "Two individuals, one focused on mere survival and the other on reproductive imperatives, neither with much chance of success outside the existing territorial packs."

However.

The map of wolf territories shifted. They fought over borders. They staged coups. Fortunes rose and fell. Packs gathered strength, then disappeared.

When Ferocious Warrior recovered from her injuries, she and Prince Charming founded the Chippewa Harbor Pack on the east end of Isle Royale. She had a scarred nose, a scarred throat, and a broken canine tooth, but she and her mate raised litter after litter of pups and killed two alpha males of the East Pack, one after the other, as they ate into East Pack territory, eventually controlling a fourth of the island.

But by the winter of 2006–2007, the Chippewa Harbor Pack was weakening. The East Pack ambushed and killed Prince Charming as he ate a moose calf on the territory border between McCargo Cove and Chickenbone Lake. As Vucetich described it, after making a beeline toward the dead moose through thick snow, then stopping to howl, the East Pack lunged at the male in a clump of trees: "In less than three minutes, [the East Pack] left only a lifeless carcass and blood in the snow," he wrote in that year's annual report. Ferocious Warrior lived for another year and gave birth to more pups, but the East Pack eventually killed her, too. The researchers found her skull and radio collar at Angleworm Lake.

The same year, Old Gray Guy died. His body was never found, leaving behind only his genetic legacy—in every single wolf on the island.

Just as they hadn't seen his death, the biologists hadn't seen Old Gray Guy's arrival. They still thought the island was a closed system with no wolves coming in or out. They still puzzled over unexplained population leaps and drops. They were in the habit of collecting whatever they could get their hands on. This included wolf scat, which they kept for more than a decade, waiting for funds for DNA tests. After they finally had the money and commissioned the tests, Peterson, on a rare mainland visit in July 2010, got a call from DNA expert Jennifer Adams requesting he come to her office. The results from the tests glowed on her computer screen.

"Sit down," she said.

The DNA tests showed that Isle Royale was not an isolated lab at all. It was just another part of the messy, complex world. Fresh genes had come in with Old Gray Guy in 1997 and possibly in the late 1960s as well, triggering the wild population fluctuations Peterson had witnessed. But Old Gray Guy chose his daughter as a second mate, and then she chose her son. Inbreeding depression—which scientists thought wolves avoided—haunted the Isle Royale population. As relatives mated, negative traits carried on recessive genes had an opportunity to express themselves. The parvovirus outbreak worsened matters, severely limiting the gene pool, a situation eased only briefly by the arrival of the virile new male (Old Gray Guy). Looking back through decades of field notes, Peterson saw them in a different light: the black wolves that skulked through the island when he started on the study; the fact that in the 2000s, a number of males turned white as they aged —these coat colors were genetic traits brought in by newcomers. He'd been looking for answers in weather and pack dynamics, but genes were the unseen engine of it all.

The Isle Royale study, over its lifetime, has shown all kinds of unexpected things, including information about the effect of diet on arthritis and evidence of a decline in mercury pollution documented through moose teeth. But when scientists were finally able to track genes, it was as though a map was pulled back to reveal another map underneath, one with unimagined mountains and strange rivers. The DNA tests uncovered the power of chance events (a tourist ignoring the rules and bringing his sick dog to the island) and individual personalities (the dominance of Old Gray Guy, the resilience of Ferocious Warrior) to shape populations. It undermined the notion of a natural balance.

On the bulletin board at Isle Royale's Daisy Farm campground, a hand-scrawled sign announces a talk about the moose and wolves of the island. It's a chilly evening, and eight people—this early in the season, mostly park employees—sit on stumps or in the tall grass. Carolyn Peterson, a longtime field assistant married to Rolf, unpacks a collection of bones. She used to stress that wolves don't suffer from inbreeding depression. Now she holds up a moose jaw and says, "Things are much more complicated than we thought . . . Things are changing all the time. We don't use the word *balance* anymore."

And that is the conclusion many researchers are drawing as models falter. A recent article in *Nature* argued that the concept of trophic cascades is flawed in its simplicity. Studies at Oregon State University are showing that the idea of a keystone species, a lynchpin for an ecosystem, may also be too basic: other species and environmental conditions play essential roles.

In a paper evaluating the case for trophic cascades in the *Annual Review of Ecology, Evolution, and Systematics*, Peterson, Vucetich, and Douglas Smith, who trained on Isle Royale and now is a project leader for the Wolf Restoration Project at Yellowstone, argue that ecosystems are too complex to trace neat relationships, particularly in Yellowstone where grizzly bears, black bears, cougars, and wolves eat bison, deer, and elk. They also point out that, when you follow the threads of prey fluctuations, you often find at the source not wild-animal predators but human beings. We are part of these ecosystems too. We decide to kill off bison, decide to bring them back, conclude there are too few elk, then too many. Perhaps we are always, forever, just studying ourselves, like Pooh and Piglet tracking the Woozle around and around a clump of larch.

There's no doubt that wolves and other large predators have an effect on ecosystems—that effect is just growing harder to graph. And there's no doubt that some questions are dangerous to ask. What are the conservation implications of questioning an idea that made such a reasoned case for wolves' importance? Peterson, Vucetich, and Smith remind their readers that, whatever their conclusions about trophic cascades, "the answer is also a weapon."

It's difficult for agencies to make management decisions without a sense of the way things will unfold, and one of the main lessons of Isle Royale is that the future is unpredictable. What if, for example, you had a highly inbred wolf population, developing extra vertebrae and bad eyes, failing to reproduce? And what if the moose population were spiraling upward, chewing trees to bits, just at the time that balsam fir might have recovered enough to drop the first seeds in a century? Should you bring more wolves from the mainland and stage a genetic rescue? No model can say.

In February 2015, an ice bridge from Isle Royale reached the mainland. Looking for the island's remaining wolves in a wind-buffeted plane, John Vucetich glimpsed something out of the corner of his eye. On the western side of the island, on a hillside near Mud Lake, two wolves slept curled against the fierce gusts. One, a solid female with brown and tan fur, patchy with mange, wore a radio collar. The other, slender and long legged, had lighter fur, almost cream-colored toward the feet. They were strangers to the island.

The two had been sniffing around for three days. The collared one liked to roam. Before her radio signal died, biologists with the Grand Portage Band of Lake Superior Chippewa charted her explorations south toward Duluth and west to Voyageurs National Park. Whatever she and her companion smelled at Isle Royale—maybe a lodge of fat beavers or a scrap of ancient moose hide—it wasn't many wolves.

Three years prior, three of the island's wolves had drowned in an abandoned mine shaft. A park biologist found their bodies floating in the pit. The carcasses included a vigorous male, alpha of the Chippewa Harbor Pack, and a young female. Other wolves simply disappeared, leaving only three on the island: a couple (who were half-siblings and father and daughter) and a crippled pup unlikely to make it through the winter. Together they were the remains of the Chippewa Harbor Pack, grandchildren of Ferocious Warrior, with a healthy mix of Old Gray Guy thrown in.

The new wolves trotted south along the shore. The future of the population was right there, listening to raven calls echo off the ridges, exploring snow-buried creeks, catching a rank whiff of fox.

Natural genetic rescue, without any environmental impact statements, fighting, or government funding. But when they reached Cumberland Point, the southern tip of Isle Royale, the newcomers headed back to Grand Portage, through hills that looked like piles of broken glass and beaches of fine ice. Not long after, the ice bridge melted away.

"For whatever reason, they went home." Peterson shrugs as he tells me the story. "That's what wolves do—they do what they want."

A year and a half after the mainland wolves turned around, a moose stands up to its chest in a lake tucked against a hot slope covered with trees in pale May leaves, soaking. Desiccated wolf scat curls on the trail nearby. Another moose dunks its whole head underwater to yank up last year's water shield rooted at the bottom, vanishing for what must be a very long breath. When he resurfaces, the splash of the ungainly head breaking the surface, the water draining off antlers and through ratty, tick-filled spring fur, is the wettest sound in the world.

Nothing disturbs the young bulls this afternoon. The wolves are down to two. But maybe not for long. After delaying for years, the National Park Service recently released a draft environmental impact statement that advocated importing as many as 30 wolves in a three-year span, for the sake of the moose, for the sake of the plants. Human judgment about the value of wolves will reshape the predator-prey landscape yet again. And the future of both the wolf and moose populations will depend, in part, on who those wolves are as individuals—cautious, brave, aggressive, intelligent.

An accident in a mine; the spread of a mutant virus; a stranger coming to town; a survivor, left for dead, beating the odds and flourishing; a dynasty overthrown. We know these stories; we just don't think of them in terms of wolves. But maybe Seton was right. Maybe there's value in discussing Big Jim, Old Gray Guy, and Ferocious Warrior, in looking at the impact of specific personalities, as much as there's value in searching for universal models. Maybe looking at the impact of specific personalities will help us to understand what is actually going on.

"What if nature is a little more like human history? When the Berlin Wall fell, when the Soviet Union collapsed, no one predicted that," Vucetich says. Isle Royale now "allows us to understand the stories of individual wolves," he adds. "This is Shakespeare in the nonhuman world."

PART III

"At the Start of Every Disaster Movie Is a
Scientist Being Ignored"

Environmental Science

DOUGLAS FOX

Firestorm

FROM *High Country News*

AIRCRAFT N2UW HAS flown through all kinds of weather. The twin-propeller plane is sleek, petite, and so packed with scientific gear for studying the atmosphere that there's barely room for two passengers to squeeze into its back seats. Monitors show radar reflections, gas concentrations, and the sizes of cloud droplets. The plane has flown through tropical rainstorms in the Caribbean, through the gusting fronts of thunderheads over the Great Plains, and through turbulent downslope winds that spawn dust storms in the lee of the Sierra Nevada. But the four people on board August 29, 2016, will never forget their flight over Idaho.

The plane took off from Boise at 4 p.m. that day, veering toward the Salmon River Mountains, 40 miles northeast. There, the Pioneer Fire had devoured 29,000 acres and rolled 10 miles up Clear Creek Canyon in just a few hours. Its 100-foot flames leaned hungrily into the slope as they surged uphill in erratic bursts and ignited entire stands of trees at once.

But to David Kingsmill, in the plane's front passenger seat, the flames on the ground two miles below were almost invisible—dwarfed by the dark thing that towered above. The fire's plume of gray smoke billowed 35,000 feet into the sky, punching into the stratosphere with such force that a downy white pileus cloud coalesced on its underside like a bruise. The plume rotated slowly, seeming to pulse of its own volition, like a chthonic spirit rising over the ashes of the forest that no longer imprisoned it. "It looked," says Kingsmill, "like a nuclear bomb."

Undaunted, Kingsmill and the pilot decided to do what no research aircraft had done: fly directly through the plume.

Orange haze closed around them, then darkened to black, blotting out the world. Kingsmill felt his seat press hard against his back as the plane lifted suddenly, like a leaf in the wind. Then the black turned back to orange. The plane jolted and fell. Pens, cameras, and notebooks leaped into the air and clattered against windows. A technician slammed headlong into the ceiling. A moment later, N2UW glided back into daylight.

According to the plane's instruments, it had been seized by an 80 mile per hour updraft of hot, buoyant air, followed by a turbulent downdraft. It was "the strongest updraft I've ever flown through," says Kingsmill, a precipitation and radar scientist at the University of Colorado in Boulder. Even stronger forces were at work several thousand feet below: the plane's radar waves, reflecting off rising smoke particles, had registered updrafts exceeding 100 mph.

Hundreds of miles away, Kingsmill's research partner, Craig Clements, a fire meteorologist at San José State University, watched the plane's flight path creep across a map on his laptop screen. The unfolding drama offered a tantalizingly detailed glimpse into the anatomy of an extreme wildfire. "It's amazing," says Clements. "We've never seen this kind of structure in a fire plume, ever." For decades, scientists have focused on the ways that topography and fuels, such as the trees, grass, or houses consumed by flames, shape fire behavior, in part because these things can be studied even when a fire isn't burning. But this line of inquiry has offered only partial answers to why certain blazes, like the Pioneer Fire, lash out in dangerous and unexpected ways—a problem magnified by severe drought, heat, and decades of fire suppression.

A mere 1 percent of wildfires account for roughly 90 percent of the land burned each year in the western United States. Some of these fires "really are unprecedented," says Mark Finney, of the U.S. Forest Service's Missoula Fire Sciences Laboratory. Their behavior "is particularly threatening because we don't have a good way to anticipate or predict [it]." So Finney, Clements, and a handful of others are increasingly turning their gaze to fire's invisible and diaphanous incarnations: the hot, roiling gases and smoke swirling among the flames, and the rising plumes they coalesce to form.

There, they believe, lies the key to understanding the way a wildfire breathes—roaring into conflagration with bigger gulps of oxygen or sputtering along more slowly on little sips. How it moves, spawning lethal fireballs or hurling burning logs ahead of the flames. The way it grapples with the upper layers of the atmosphere, sending embers in unexpected directions to propagate itself across the land. Even, perhaps, the role its elemental opposite —water—plays in driving its explosive growth.

Nailing those connections could provide new tools for monitoring fires and predicting their behavior. This could give firefighters precious minutes of advance warning before potential catastrophes, and better inform the difficult decision to order an evacuation.

But it won't be easy. "The plume is orders of magnitude harder to study than the stuff on the ground," says Brian Potter, a meteorologist with the Pacific Wildland Fire Sciences Laboratory in Seattle who sometimes works with Clements. Indeed, it took a global conflagration much darker than any forest fire to even begin laying the foundations of this work. Kingsmill's observation about the bomb, it turns out, isn't far off.

The evening of July 27, 1943, was stiflingly hot in Hamburg, Germany. The leaves of oak and poplar trees hung still in the air as women and teenagers finished factory shifts and boarded streetcars. They returned home to six-story flats that lined the narrow streets of the city's working-class neighborhoods. They opened windows to let in cooler air, and folded themselves into bed. It was nearly 1 a.m. when British planes arrived.

Searing yellow flares drifted down over the city, dropped to mark the city's eastern quadrant as that night's target. Bombs followed, tearing open buildings and exposing their flammable contents to a rain of incendiary canisters that hissed as they fell.

Thousands of small fires sprang up. Families retreated into basements. The buildings above them roared into flame, and these growing fires greedily sucked air from their surroundings. Their collective inhalation drew winds through the narrow urban canyons, pulling along embers that ignited yet other buildings. Within minutes, the fires were merging.

The British historian Martin Middlebrook has collected the accounts of survivors who dampened their clothing, fled their bunkers, and crawled through the streets. They described the inflowing winds as "shrill," "shrieking," and "howling"—the scream of "an old organ in a church when someone is playing all the notes at once." Gales exceeding 110 mph uprooted trees, pushed struggling full-grown adults deeper into the fire zone, and sucked babies and elders into burning buildings.

The winds swirled into flaming tornadoes that swept up people and turned them into "human torches." Balls of fire shot out of buildings. Within 60 minutes, a spiraling pillar of smoke had swelled into an anvil-shaped thunderhead that towered 30,000 feet over the city.

At least 42,000 people died, and another 37,000 were injured. Sometime during the night, someone scribbled a word for the unspeakable destruction into the logbook of the Hamburg fire department: "feuersturm"—in English, "firestorm."

Such cataclysms had occurred before. Fires destroyed London in 1666; Peshtigo, Wisconsin, in 1871; and San Francisco following the 1906 earthquake. But Hamburg might have been the first time that people intentionally created a firestorm, with chilling calculation.

The British chose to bomb that section of the city, not just to demoralize the workers in Germany's critical U-boat industry but also because of the tightly packed buildings that covered 45 percent of its ground area. And their tactics were almost certainly influenced by experiments begun four months earlier, across an ocean and half a continent, on a remote desert playa in northwest Utah.

There, the U.S. Army's Chemical Warfare Service had commissioned Standard Oil Development Company to construct a row of steep-roofed European-style apartment buildings. Erich Mendelsohn, an architect who had fled Nazi Germany, specified every detail: 1¼-by-2-inch wood battens, spaced 5⅞ inches apart, to hold the roof tiles; 1-inch wood flooring underlain by 3½-inch cinderblocks, and so on—all to replicate the dwellings of German industrial workers. The wood was maintained at 10 percent moisture to mimic the German climate. Rooms were outfitted with authentic German curtains, cabinets, dressers, beds, and cribs—complete with bedding—laid out in traditional floor plans.

Then, military planes dropped various combinations of charges on the buildings, seeking the most efficient way to penetrate the roofs and lace the structures with flame.

Those experiments offered clues on what factors could cause firestorms. And in the years following World War II, scientists would study Hamburg and other bombing raids to derive basic numbers for predicting when a firestorm might form: the tons of munitions dropped per square mile, the number of fires ignited per square mile, and the minimum area that must burn. They concluded that Hamburg's unusually hot weather set the stage for the firestorm, by making the atmospheric layers above the city more unstable and thus easier for a smoke plume to punch through. Scientists theorized that this powerful rise had drawn in the winds that whipped the flames into even greater fury.

Later, scientists studying urban fires during the Cold War noticed something that underscored this finding: when a fire plume rotated, the rate of burning seemed to increase on the ground. It suggested that rotation lessened the drag between the plume and its surrounding air, allowing it to rise more strongly and pull in fresh oxygen more effectively on the ground.

Fire is so universally familiar that we take for granted that we understand how it works. And yet these old experiments, finished by 1970, are still a key source of knowledge about extreme fire behaviors. Until recently, technology was simply too limited to reveal much more about the specific mechanisms by which a fire plume might feed a firestorm, let alone how beasts like fire tornadoes and fireballs form. Scientists needed new ways to see within the smoke and turbulent flames—to make the invisible visible. And as it happens, they began finding them almost by accident.

One morning in February 2005, Craig Clements watched as a six-foot wall of flame crept across a prairie a few miles outside Galveston, Texas. He was not yet a fire scientist; in fact, he was slogging through his seventh year of graduate school, studying a completely unrelated topic—mountain winds. This was just a side project, a favor he was doing for his Ph.D. adviser at the University of Houston, who had a steel weather tower in the field. A prescribed fire had been planned there to prevent fuel buildup that could cause a more serious blaze. What would happen, they wondered, if they mounted extra instruments on the tower to measure the winds, heat, and gases released by the fire passing directly beneath it? The results were stellar: Clements's sensors showed that the flames produced a surprisingly strong pulse of water vapor.

Scientists knew, theoretically, that the combustion of dry plant matter would release water vapor along with carbon dioxide, but it had never been measured this carefully in a real fire. And when Clements showed his work at a conference several months later, scientists there implored him to do more experiments. "It rescued my life, my academic career," Clements now says. "Totally."

Clements abandoned his previous line of study and landed a faculty job at San José State in 2007, where he continued his research. His instrument towers, deployed in carefully controlled fires, provided yet more unprecedented and precise measurements: how winds accelerate and draft into an advancing flame front, the heat and turbulence above the flames, and the speed of the rising hot air.

Still, dangerous behaviors like fire tornadoes, or lofting of embers, usually happen in much larger fires than experiments can replicate. And the plumes of those fires rise thousands of feet. Clements was capturing the action only within a few feet of the ground. And he was just getting point measurements—like following a single bird to understand the movement of an entire flock wheeling in the sky.

Clements wanted to capture the whole phenomenon—to look inside the opaque mass of an entire fire plume from a distance, and see all of its parts swirling at once. In 2011, he found his lens: a technology called Doppler lidar.

Unlike the Doppler radar that police use to measure the speed of passing cars, lidar is tuned to detect reflections of its low-powered laser off particles smaller than red blood cells. It actually scans the sky, collecting thousands of pinpoint measurements per second, which can be reassembled into a picture of both the plume's surface and its internal air currents. Clements and his then–postdoc. student, Neil Lareau, mounted this television-sized gadget in the back of a pickup truck and hit the road in search of wildfires.

In June 2014, live ammunition fired during an army training exercise afforded them the chance to watch a fire roll through 4,800 acres of grass and oak hills at the Fort Hunter Liggett training ground in California. They watched through lidar as a rotating column of smoke stretched, narrowed, and accelerated into a fire tornado two football fields across, with winds swirling 30 mph.

These tornadoes, or "whirls," can pose sudden dangers in wildfires. During a 1989 blaze near Susanville, California, a powerful whirl raced out of a flame front, with winds estimated at 100 mph. Three fire engines retreated just in time to escape being torched, but four crew members were hurled into the air—all of them seriously burned.

For now, fire whirls are nearly impossible to predict. But that afternoon at Fort Hunter Liggett, Clements and Lareau began to get a sense of how they form.

It was an eerie and beautiful process, hidden deep inside the smoke column. First, an embryonic disturbance in the fabric of the plume: hot rising gases began to rotate with the motion of an air current coming from the side. This vague motion coalesced into two small, separate whirls. They circled around one another like dancing, swaying cobras preparing to mate, then merged into a single powerful vortex.

"The laser is seeing through a lot of the smoke," says Lareau. "It's showing you something that you can't necessarily see by the naked eye about what the fire is doing." If firefighters had access to similar technology, they could potentially recognize an impending whirl in real time before it forms, and escape.

A month after Fort Hunter Liggett, Clements and Lareau stumbled onto another discovery at the 17,000-acre and-growing Bald Fire, north of Lassen Volcanic National Park in California. On a warm, hazy morning, the pair sped in their pickup truck along Highway 44, south of the fire, looking for a good vantage point for seeing the plume. From the bed of the truck, their lidar and radar wind profiler pointed straight up at the sky, recording the smoke, winds, and clouds directly overhead. Then the highway began to gradually descend down the slope of the volcanic plateau where the fire burned, and they noticed something strange. Even as the winds thousands of feet up blew north, the smoke just below those winds was drifting steadily south.

They spent the rest of the day following the broad mass of smoke as it oozed 20 miles downhill, like a gauzy, viscous lava flow in the sky. During that drive, they downloaded weather and satellite images of the broader smoke plume from both the Bald Fire and the similar-sized Eiler Fire, 10 miles to the southwest of it. A surprising picture emerged.

Smoke from the Bald Fire had shaded broad swaths of the landscape—cooling the ground, and several thousand feet of air above it, by a few degrees. Even as the winds blew north, this cool, dense air was rolling downhill like molasses, pushing under the winds as it followed the contours of the land—carrying a layer of smoke 6,000 feet thick along with it.

"We had no idea we were going to see things like that," says Lareau, who now holds a faculty position at San José State. "It seems like every time we go, we end up with new perspectives." The team's insight about the Bald and Eiler Fires has implications for predicting smoke and air quality—a constant concern for communities near large fires. It also impacted the fires themselves. Even though both fires existed in the same atmospheric environment of pressures and winds, and burned across similar terrain, they were spreading in opposite directions that day—Bald to the south and Eiler to the north. This denser current of cold air and smoke was actually pulling the Bald Fire in the opposite direction of what was predicted based on wind alone.

Clements imagines a future in which lidar is not simply a tool of research but also standard equipment mounted with automatic weather stations on fire trucks.

This device would theoretically be much smaller and cheaper than current technology. It scans the plume continuously to obtain real-time data, which "is then uploaded into a mainframe computer that's running a fire-weather model," says Clements, "and boom, problem solved." The fire crew would receive a fire-behavior forecast that reflects detailed information about a plume's evolving structure—something not currently possible. That forecast could warn about impending events, such as a strengthening updraft that might conspire with winds higher up to toss embers into unburned areas, or an incipient plume collapse that might splash the fire and hot gases in unexpected directions.

Yet even in its ideal form, that bit of technology wouldn't be able to forecast everything: some extreme fire behaviors are driven by smaller-scale forces that even lidar can't capture. Scientists suspect these might be responsible for some of the more tragic firefighter deaths in recent years.

The South Canyon Fire hardly seemed threatening at first. Lightning started it on July 2, 1994, atop a ridge overlooking Interstate 70, a few miles west of Glenwood Springs, Colorado. It crept at a civil pace down the mountain's slopes, through dry grass and Gambel oak. Then, around 3 p.m. on July 6, a cold front swept over the area, spawning winds that pushed the lower part of the fire across the mountain's southern face—igniting the base of an unburned drainage. The fire crew working there may not have realized their peril until too late.

Fourteen firefighters were crossing the drainage as the fire entered its lower reaches. The flames quickly gained on them as they hurried diagonally across it, toward the ridge top, following a firebreak through the dense vegetation that they had cleared the day before. Firefighter Kevin Erickson, a couple hundred feet in front of the others, glanced back to see a wall of flame advancing up the sides of the gully before he crested the ridge and scrambled down its other side. Firefighter Eric Hipke was 45 seconds behind him. His pace quickened as the heat grew unbearable. Several steps short of the ridge top, a blast of searing air struck him from behind. He slammed to the ground with a yell, then scurried to his feet, shielding his face from a maelstrom of smoke and flying embers, and sprinted over the top.

Hipke was the last person to reach the ridge line alive. Only later did he realize that his backpack's shoulder straps had melted through, leaving it bobbing from his waist belt. He sustained third-degree burns across the back of his arms, legs, torso, and head. The bodies of the 12 remaining crew members were found that evening. (Two from another party, who died in a separate part of the blaze at the same time, were found July 8.) Hipke's crew was strung out along the path they'd been following, some with their backpacks still on—as though overcome, simultaneously, by a sudden force.

Observers have speculated for years about what, exactly, killed them. The fire may have overtaken them or a gust of wind may have pushed the hot gases of the plume down onto them. In the years since, however, scientists have uncovered another possibility —a type of blowup that may have caused multiple fatalities over the years but left no survivors to describe what happened.

Like Clements, Janice Coen stumbled onto these questions by accident. Coen works at the National Center for Atmospheric Research in Boulder, Colorado, where she studies fire's inner workings. In September 1998, she spent several hours aboard a Hercules C-130 aircraft as it circled over Glacier National Park. The McDonald Creek Fire was marching up a steep slope at roughly three feet per second. Its smoke obscured the advancing flames,

but infrared video cameras mounted outside the plane recorded
what was happening underneath. It was only later, as Coen looked
through individual frames of that video, that she noticed something strange: at one point, a jet of flame seemed to shoot ahead
of the fire. It lasted only a second or two but left a trail of newly
ignited vegetation in front of the fire. Not until Coen calculated
the size of the pixels and the time between frames could she appreciate its true significance.

The jet had surged 100 yards ahead of the fire's front, advancing 100 mph—"like a flamethrower," she says. It was 10 times faster
than the local wind—generated, somehow, by the fire's own internal tumult.

Coen called it the "finger of death," and for her it brought to
mind the unconfirmed reports of fireballs that occasionally circu-
lated among firefighters. She had never seen such a thing, but
as she examined footage of other fires, she was surprised to find
fire jets again and again. Her infrared videos were in some ways
akin to those classic blurry clips of Sasquatch walking in a forest
—a strange and fleeting embodiment of fire's turbulence, without
clear explanation for its existence.

When you think about turbulence, what comes to mind is something felt but not seen—like a bumpy airplane ride, where the air
currents themselves are invisible. Only in the special case of fire
is it possible to see turbulence with the naked eye—sort of. The
flames are composed of hot, glowing gases; their flickers and licks
are the roiling movements of those gases. That movement unfolds
too quickly for the eye to comprehend. So Finney has spent years
slowing down videos of fire in experiments at his Missoula laboratory, rewinding and replaying them, exposing the secret details
that have long hidden in plain view. An advancing flame front
seems chaotic. "But there is organization in there," he says. There
are "flame structures that [are] very repeatable."

In his lab experiments, Finney used high-speed cameras to
watch what amounted to miniature forest fires: walls of flame advancing through hundreds of cardboard "trees" the size of match-
sticks.

The advancing flame front resembles a jagged-toothed saw
blade at any given instant, with interspersed high and low points,
flickering several times per second. But slow it down, watch a single one of these high points, and you begin to perceive something

more complex. The flickering peak of flame repeatedly curls over on its side, like a surfing wave in Hawaii, viewed edge-on as it rolls into a pipe and crashes on itself.

This churning wave of flame rolls over and over, staying in roughly the same place. It is a horizontal, rolling current, driven by the constant push and pull of gases within the fire. Combustion gases heated to 1,800 degrees Fahrenheit are only one-quarter as dense as ambient air—a difference that makes them more buoyant and causes them to rise, creating the flickering peaks of flame. Cooler, denser gases rush downward to fill the void, driving the downward side of the rolling current and pressing down on the fire to create low spots in the flame front.

Finney's slow-motion videos show that these rolling eddies exist in pairs within the fire. They roll in opposite directions, coupled like interlocking gears. Their combined motion periodically pushes down on the advancing front of the fire, causing flames to lick downward and forward, ahead of the fire.

Finney believes that these forward flame-licks are scaled-down versions of the "fingers of death" that Coen has seen in wildfires—possibly even related to the fireballs said to have shot out of buildings during the 1943 Hamburg firestorm.

Coen has actually documented similar flame-rollers in real wild fires using infrared video. But she believes that the finger of death also requires another factor. As bushes and trees are heated by an approaching fire, their decomposing cellulose releases hydrogen, methane, carbon monoxide, and other flammable gases in a process called pyrolysis.

Coen and Shankar Mahalingam, a fluid-dynamics engineer at the University of Alabama in Huntsville, believe that rolling currents can mix these flammable gases with oxygen-rich air. "The dangerous situation is when the fire is going up on a hill," says Mahalingam. "Maybe there are pyrolysis products that have accumulated" in front of the fire and mixed with fire-boosting oxygen. As the flame licks forward into this invisible tinderbox, it ignites a blowtorch.

Reflecting back on the South Canyon Fire in 1994, it is tempting to wonder whether the same blast that flattened Hipke near the ridge top also killed the 12 firefighters behind him—one of them just 40 yards back. Some have even speculated that Hipke would have died along with them—his lungs seared by hot combustion gases—had he inhaled rather than yelled as he fell to the ground. Coen sees the South Canyon tragedy as one of several likely caused by the "finger of death"—a monster created by the turbulent respiration of the fire itself and the violent rise of its hot, buoyant gases.

These same buoyant gases also supply the momentum that drives a fire whirl to spin once it is triggered. And on a much larger scale, they are what pushes a fire plume ever higher in the sky, powering the in-drafts that keep the fire burning below.

But the source of the speed and energy with which these gases rise is still the subject of intense speculation. Potter, of the Pacific Wildland Fire Sciences Laboratory, has found some surprising possible answers. They arise, in part, from some of those old military fire experiments—these particular ones conducted in the aftermath of World War II by a U.S. government that feared the devastation of Hamburg might represent the future of modern warfare.

In the wrinkled, sage-covered mountains of Nevada near the California border, 30 miles east of Mono Lake, there is a meadow that seems to lie in shadow even on sunny days. Spread across it are hundreds of dark patches, where the soil is mixed with charcoal. These spots lay row upon row, like the ghostly foundations of a dead city. In a sense, that is exactly what they are.

In 1967, workers with the Forest Service and the Department of Defense stacked 342 piles of juniper and piñon logs in this place —20 tons of wood per pile, spaced 25 feet apart. Then, at 8 a.m. on September 29, they set fire to them.

Project Flambeau comprised some two dozen experiments like this one, meant to simulate an American suburb under nuclear attack—specifically, the many small fires that would merge into a storm, as happened not only in Hamburg but also in Hiroshima and Nagasaki.

Lengthy reports describe how helium balloons released near the fire here rose several hundred feet, then swooped down into the flames, revealing strong downdrafts feeding the fire from its sides. But what drew Potter's interest was the water. Concentrations of water vapor rose 10 to 20 times higher than the surrounding air.

Water is a major product of combustion, second only to carbon dioxide. It forms as oxygen binds to the hydrogen atoms in wood, gasoline, or just about any other fuel—creating hydrogen oxide, otherwise known as H_2O. Burning four pounds of perfectly dry wood releases a pound or two of water.

Exhale onto a car window and you will see another form of the same phenomenon, fogging the glass: water produced from the oxidation of food you have eaten. This vapor is familiar and mundane; it hardly seems like a violent force.

And yet water vapor fuels the strongest updrafts in nature, says Potter, from thunderstorms to tornadoes to hurricanes. As moist air rises during these storms, the water vapor condenses into cloud droplets, releasing a small amount of heat that keeps the air slightly warmer than its surroundings, so it continues to rise. "Water," he says, "is the difference between a weak updraft and a really powerful updraft."

Potter wondered if the water vapor released from combustion might infuse extra energy into wildfire plumes. By condensing and giving off heat, it might allow some plumes to rise higher and faster, accelerating the fire on the ground.

Through a bit of serendipity, this theory actually led to Clements's first fire experiment—the prescribed prairie burn back in 2005. It was Potter, who happened to know Clements's Ph.D. adviser, who suggested it. Neither the results of Clements's experiment nor those of Flambeau were conclusive about the importance of this pulse of water vapor. Still, some people have latched onto the theory.

Michael Reeder, a meteorologist at Monash University in Australia, is one of them. He believes that water was pivotal in fueling the firestorm that swept through the suburbs of Canberra, the Australian capital, on January 18, 2003.

The fire consumed 200,000 acres of drought-stricken territory that day, isolating the city under a glowing haze of Halloween orange. Remote infrared scans suggest that during a single 10-minute period, it released heat equivalent to 22,000 tons of TNT— 50 percent more than the energy unleashed by the atomic bomb dropped on Hiroshima.

A series of four pyrocumulonimbus clouds rose into the stratosphere that afternoon. These fire-fueled, anvil-shaped thunderheads lofted black, sooty hail up to six miles away. One of them spawned a tornado that snapped the tops off pine trees as it plowed a path of destruction 12 miles long and a quarter mile wide.

The tornado and the height of the clouds "point to something extraordinary," says Reeder. "[They] require moisture—and the question is: how do you get that much moisture over eastern Australia during drought conditions?"

Combustion provides a plausible source for it. Reeder estimates that the fire incinerated over 2 million tons of wood and vegetation that day, liberating at least a million tons of water vapor into the sky.

The temperature and density differences that drive such cataclysmic power can seem deceptively minuscule. When N2UW flew through the plume of the Pioneer Fire in 2016, its instruments registered updrafts of 80 to 100 mph. Yet at that elevation, 8,000 feet above the flames, the interior of the plume was only 3 to 6 degrees Fahrenheit warmer than the surrounding air, meaning that its buoyant stampede through the atmosphere was powered by a density difference of just about 1 percent.

In other words, given the right atmospheric conditions, a few degrees of warmth and extra buoyancy could spell the difference between a plume that pushes 40,000 feet up, into the stratosphere, powering a vicious blaze on the ground—as Pioneer did—and one whose smoke never escapes the top of the boundary layer at 3,000 feet, leaving the fire stunted, like a weather-beaten dwarf tree gasping for life at timberline.

N2UW made two more passes through the plume of the Pioneer Fire on August 29. During that third and final pass, static electricity roared through the cockpit radio. Concerned that lightning from the plume might strike the plane, the pilot turned off his antenna.

That flight yielded far more than the first direct measurement of a plume's updraft. Days later, Clements found himself looking at a portrait of the fire's plume unlike any that has existed before: a vertical MRI slice of sorts cut along the path of the plane—captured by its fine-tuned scientific radar, aimed straight down.

Color-coded by the velocity of its air currents, the blotchy mass resembled a hovering spirit—large-headed, legless, and deformed. Clements's trained eye began to pick out some basic structures: a 40 mph downdraft next to a 60 mph updraft signified a turbulent eddy on the edge of the plume. Hot air pushing up past cooler, stationary air had set in motion a tumbling, horizontal vortex—

the sort of thing that could easily have accounted for the plane's brief freefall. Those blotchy radar pictures may finally allow us to see through wildfire's impulsive, chaotic veneer—and perceive the more predictable, underlying forces that guide its behavior. "We didn't even know this would work," said Clements. "This is the most exciting thing I think I've ever seen in my career."

Simply seeing can be transformative. Not until people saw microbes could they comprehend and fight diseases like malaria —once blamed on foul spirits or miasmas. And not until Earth's colorless, odorless magnetic field became visible could people appreciate how it shaped the planet's environment.

While the smoke plume of the Pioneer Fire was apparent to the naked eye, the violent forces within it were also deceptively invisible. As the plume first approached it on August 29, the pilot's standard weather console showed the plume as nothing but a swath of cool blue—a seemingly gentle updraft, with no hint of what lay in wait.

J. B. MACKINNON

Tragedy of the Common

FROM *Pacific Standard*

WHITE-RUMPED VULTURES WERE one of the most common
large birds on the planet. There were an estimated 40 million of
them in India alone. One famous bird sanctuary, Keoladeo National Park, had 30 vulture nests per square mile. Even in Delhi,
India's capital city and the second-largest urban agglomeration in
the world, they averaged 8 per square mile.

Biologist Vibhu Prakash found it captivating to see hordes
of the hungry scavengers feeding, their backs like a phalanx of
shields that heaved with the rhythm of beaks tugging flesh. In just
20 minutes, the birds could pick a cow carcass clean.

"They were in so big a number, villagers would be scared passing near the congregation," said Prakash, a principal scientist with
the Bombay Natural History Society. "When they are feeding on a
dead body and all, they look a little scary."

In 1997, Prakash headed into Keoladeo, a small UNESCO
World Heritage site 100 miles south of Delhi, in order to update a
count of white-rumped vultures that he had carried out a decade
earlier. He was about to become the principal witness to an accelerating vanishment that, even in this age of extinctions, retains its
power to shock.

When his tally was complete, Prakash had recorded a 58 percent drop in the Keoladeo vulture population. "I saw the dead
vultures almost everywhere," he said. "We saw dead birds inside
the park, and then we also saw dead birds outside the park." Local
villagers, who relied on vultures to keep rotting livestock carcasses
from spreading disease and sold the cleaned bones to be ground

into fertilizer, confirmed that there were far fewer of the birds around. Prakash began to investigate further, ultimately launching a 7,000-mile road survey that uncovered a 90 percent decrease in vultures nationwide.

Years of study finally determined that a veterinary drug, diclofenac, was persisting in livestock carcasses and poisoning vultures after so much as a single exposure. Though the drug is now banned, illegal use remains a problem. The white-rumped vulture die-off has now reached 99.9 percent, and the species has joined the dismal roster of more than 5,000 life-forms assessed as critically endangered by the International Union for Conservation of Nature.

Today, Prakash said, many Indians under the age of 25—roughly half the population—don't believe him when he describes the birds' recent omnipresence. "They've not actually seen a vulture," he said. "They think I'm telling stories."

Nature is like granola: the list of ingredients is long, but the bowl is mostly filled with just a few of them. Take England, for example, which is small and critter-obsessed enough to count its wildlife nearly one by one. Population estimates for 58 species of land mammals in that country, ranging from the familiar to the obscure, total about 173 million animals. But just three species—the common shrew, rabbit, and mole—account for half of those individuals. All told, the most common 25 percent of English mammal species add up to 97 percent of all the individual animals. Similar patterns play out on land and at sea, in your local park or across whole continents, and whether you're counting beetles, shellfish, or tropical trees. The most common land bird in the United States and Canada is the American robin, harbinger of spring. Robins alone are as numerous as the two countries' 277 least common bird species combined.

That species of such incredible abundance can decline as quickly as the white-rumped vulture did points to a counterintuitive idea in conservation: that common species may need protection just as much as rare ones do

The first scientist to propose the conservation of the common was, almost too perfectly, the author of a book called *Rarity*. But after 20 years of studying what made some species rare, Kevin Gaston, an ecologist at the University of Exeter, in England, started to wonder why other species are widespread and abundant. He soon came to a seemingly contradictory conclusion: "The state of being common is rare." While any given common species is made up of many individuals, only a small fraction of species are common.

Gaston's work culminated in "Common Ecology," a paper published in the journal *BioScience* in 2011 that found that commonness was not a well-studied phenomenon, and that "many common species are as poorly studied as many rare ones." The work triggered a quiet surge of research. A study from 2014 hints at the scale of what's been overlooked. Its authors found that the number of birds nesting in Europe has dropped by 421 million—fully one-fifth of the continent's bird population, gone—since 1980, and that this decline in sheer birdiness is accounted for almost entirely by common species, among them such household names as the skylark.

Industrial agriculture carries much of the blame for Europe's disappearing birds. "They've been taking out hedgerows, taking out trees, making fields bigger, increasing inputs of insecticide, pesticides—just essentially squeezing out the opportunities for wild organisms to live in those kinds of environments," Gaston told me. "We're talking just massive losses."

But even the most human-adapted and urban of birds, such as starlings and house sparrows, have steeply decreased—in fact, those two very common birds were among the top five birds experiencing population declines. Most of the rarest birds in Europe are actually increasing at present, due to successful conservation efforts, although they remain uncommon; meanwhile, most of the common birds are declining toward scarcity. "The inevitable place you end up," said Gaston, "is that everything is rare."

In the annals of extinction and near-extinction, many of the most infamous cases involved species that were once incredibly common: the plains bison, the passenger pigeon, the Carolina parakeet. The pattern continues today. Consider the radiated tortoise, one of the world's most beautiful tortoises, its carapace a geodesic dome roofed with elaborate parquet tiles.

The tortoise is capable of extraordinary plenitude—a survey from the year 2000 estimated 10 tortoises per acre in the most pristine redoubts of the species' range on the African island of Madagascar. But after millennia of coexistence with Malagasy peoples (many of whom considered eating the tortoises *fady,* or taboo), the species began to fade. Local traditions broke down, the human population shot up, and tortoise hunting and collection for the pet trade surged (one biologist followed a local transit bus as it stopped 11 times in 10 miles to let passengers capture every shuffling shellback they saw). In 2000, researchers declared the prospects for the radiated tortoise "worrisome." A decade later, the species had leaped to critically endangered status, one step away from extinction in the wild.

Other recent examples include the saiga (an Asian antelope that looks like something out of *Star Wars*, which has plummeted from almost a million to 50,000—a 95 percent drop—since the end of the Soviet era); the European horse chestnut; sharks in general; and, of course, the white-rumped vulture.

And yet, formerly common species make up only a small fraction of the living things that are threatened with near-term extinction today. The vast majority of critically endangered species are those that were relatively rare to begin with, for the simple reason that it's much easier to drive a species with low numbers or very limited distribution to the brink.

To focus only on that final moment of total extinction, however, is to downplay the breadth of the extinction crisis. A far more frequent occurrence is *extirpation*, or local extinction—the disappearance of a species from one or another place where it used to live. Tigers, for example, have vanished from 93 percent (and counting) of their former habitat, including 10 entire nations in Asia. The Pashford pot beetle, on the other hand, is thought to be extinct but was never known to live beyond certain bogs in east-central England. One species is gone but was never abundant in the first place; the other is still with us but has experienced staggering losses.

The authors of one recent study found that the rate of population loss among terrestrial vertebrates is extremely high, even in "species of low concern." They wrote that "beyond global species extinctions Earth is experiencing a huge episode of population declines and extirpations" and used the term "biological annihilation" to describe the magnitude of the crisis. Remarkably, they characterized the wave of local extirpations as a "much more serious and rapid" decline than mass extinctions.

The current estimate among population biologists is that the planet has lost half the individual animals, plants, and other living things that make up our visible world. Most of these accumulated deaths have come at the expense of common species. They are the animals killed most often by hunters, the creatures most likely to end up as roadkill, the trees and plants that die in large numbers each time land is cleared for a farm or a housing development. Even birdwatchers dismiss the most familiar species as "dirt birds." The term derives from the saying "common as dirt."

Widespread and abundant species are often seen as natural resources, like copper or oil, rather than as living things. Of all the fish in the oceans, just 10 species account for almost a third of the global catch—Alaska pollock, chub mackerel, Atlantic herring, yellowfin tuna, and Japanese anchovy among them. The United Nations' Food and Agriculture Organization classifies every one of these fish stocks as either "fully fished" or "overfished." Similarly, the 10 most common tree species in any given nation will, on average, provide 76 percent of the wood and pulp production. In the United States, home to sprawling forests and more than a thousand native tree species, just three—Douglas fir, loblolly pine, and western hemlock—account for one-quarter of the timber harvest. It has been the lot of common species to be mistakenly thought of as infinite in number. Until, one day, it turns out that they aren't.

Despite all this doom and destruction, even protectors of the environment have tended to overlook the plight of common species. Conservation biology is forever scrambling to pull yet another species back from the void. One result has been the paradox witnessed among nesting birds in Europe, where rarer species are generally increasing while common species have declined by the tens of millions.

Two humble species of newt illustrate how this kind of situation can arise. The great crested newt, which looks like a tiny dragon carrying fire in its belly, is endangered in Belgium. The smooth newt, itself decorated with polka dots and a mini dinosaur crest, is a common species.

Researchers looked at 74 farmland ponds in Belgium; crested newts turned up in just 12 of them, while smooth newts were found in 33. One way to save the endangered crested newt is obvious: protect its dozen ponds. Yet if the remaining ponds are polluted, drained, or otherwise destroyed, then eventually the smooth newt, too, will have only 12 ponds to dwell in. That trend again: *everything ends up rare.*

To conserve the commonness of smooth newts, you must preserve a larger number of ponds, which will have the parallel effect of conserving the crested newt as well. All of which sounds manageable when you're talking about a scattering of water holes. But setting aside habitat to save a species no longer works at the scale of, say, a continent.

Take wolves, which were once the most widespread large mammals on Earth, having roamed from Japan to India to Ireland and from Mexico to the high Arctic. In the continental United States, the only place you couldn't find wolves was the sliver of California west of the Sierra Nevada.

But by the time the Endangered Species Act was signed into law in 1973, wolves were among the first animals needing protection — they had been all but eradicated in the Lower 48. Red wolves today survive only in a postage-stamp parcel of North Carolina. But gray wolves, thanks to protection and reintroduction efforts, have sprung back: 5,550 of them now roam in the western Great Lakes states, the northern Rockies, small patches of Arizona and New Mexico, and Washington State near the border with Canada.

The gray wolf is no longer at immediate risk of extinction in the Lower 48. With a relatively stable population now occupying 10 percent of its former range, the federal government has gone so far as to argue that the species has "recovered." Critics have called this interpretation of the Endangered Species Act a "museum-piece approach," because it retains species as cloistered artifacts in the smallest possible areas in which they can survive. But much of this is an accident of design. The act was created to keep rare species from going extinct; it was never up to the task of preventing species from becoming rare in the first place.

By some estimates, there is still enough habitat available to see wolves howling again in 31 states, among them California, Texas, Kentucky, New York, and Alabama. Yet it's impossible to bring back common species, or to preserve what commonness remains, through the traditional approach of setting aside protected areas — the whole planet would soon be a park. As one wolf researcher put it, "Society has not yet answered the question of how much of the landscape ought to be shared with the nonhuman world."

"The maintenance of commonness presents a different set of challenges," Gaston said. "It's much more about what you're doing across entire landscapes. How environmentally friendly is your farming? How sustainable are your fisheries?" The conservation of the common represents a deeper ambition than the twentieth century's lopsided division of the world into islands of wild, protected places in a sea of ruinous human activity. It calls on us to integrate conservation into every aspect of human life.

In the age of Donald Trump, however, such high-mindedness is upstaged by more practical questions: Is it wise to point out inadequacies in the Endangered Species Act when the act itself is imperiled? Is it sane to worry about commonness while species after species is frog-marched into actual oblivion? On the day of his inauguration, the president put a hold on a U.S. Fish and Wildlife Service ruling that the rusty-patched bumblebee be listed as an endangered species.

In the end, the bee prevailed—at least on paper. Its listing became effective on March 21, 2017. Though it's not yet clear what actions the federal government will take to save the species, the case was too clear-cut to dismiss out of hand. Rusty-patched bumblebees now occupy just 0.1 percent of their historical range, and have declined in number by an estimated 93 percent. The causes are exactly the kind of large-scale, modern-life challenges that Gaston talks about: pesticide use, the conversion of grasslands into industrial farms, climate change, novel diseases.

One other point that should not go unmentioned: little more than two decades ago, rusty-patched bumblebees still buzzed in 28 states. Residents of Midwestern cities remember shooing them out of the way as they walked the streets. The bee was, according to one Fish and Wildlife Service report, "so ordinary that it went almost unnoticed as it moved from flower to flower." In other words, America's newest endangered species used to be common.

It is the curse of today's biologists that they are forever being asked to justify the continued existence of nonhuman life. Every species, common or rare, is the product of millions of years of evolution, its journey to this point in time exactly as miraculous as our own, but this is evidently not justification enough. Can you eat it? Does it suck carbon out of the atmosphere? Could it cure cancer? If you squeeze it hard enough, will it ooze oil that can run my car?

The defense of rare species on such grounds is often difficult. The extinction of one or another rarity—rest in peace Perrin's cave beetle, hazel pigtoe, hairy wikstroemia—often comes at no obvious cost to humans or the larger landscape. I'm not saying that these plants and animals matter only to themselves. Taken together, rare species are the fail-safes, special teams, niche players, and understudies of the living world—today's rarity could, with a sudden shift in evolutionary pressures, be tomorrow's common species. Yet the loss of any single rare species amounts mainly to an incremental waning in originality, redundancy, and specialization in the environment. A wound, but largely an invisible one.

Now consider the consequences of losing a common species. Atlantic herring gather every autumn on the northern flank of the Gulf of Maine in spawning shoals that, if you could drag one ashore, would bury the whole of Manhattan six stories deep in sparkling silver fish. We know this because scientists, hailing mainly from various Massachusetts research institutions, set out in 2006 to measure them by bouncing sound waves off those giant schools. Through their technology, they witnessed what they described at the time as "perhaps the largest massing of animals ever instantaneously imaged in nature." In a study published last year, the researchers referred to the spawning event as a "massive ecological hotspot"—though the spot wasn't a place so much as a species.

When they got back to the lab and analyzed their audio recordings, the researchers discovered something equally stunning: a chorus of underwater songs, cries, down-sweeping "meows," whistles, creaks, buzzing, clicking, chirping—tens of thousands of sounds, many of them outside the range of human hearing. They were the vocalizations (the voices, we might even say) of whales and dolphins.

It was a vast assembly of marine giants, from fin and humpback whales that plowed with gaping mouths through the herring schools by night, to sperm and killer whales gorging on the fish by day. The scientists were able to identify 10 of the whale and dolphin species; 4 of them are endangered throughout U.S. waters. Other herring predators include sharks, rockfish, dogfish, bluefish, hake, pollock, seals, sea lions, tuna, and an enormous variety of seabirds—it's almost easier to draft a list of what doesn't eat herring than a list of what does. Even squid eat herring—and so do we. Atlantic herring is currently our sixth most important fish stock.

It wasn't always this way. In 1977, the herring population had collapsed—plummeting to 10 percent of its historical size—due to overfishing. Today, the Gulf of Maine herring are a rare good news story, once again as numerous as they were half a century ago. Michael Jech, a biologist with the Northeast Fisheries Science Center who was a part of the recent herring study, characterized that abundance with the understatement of a government scientist: "It seemed to be quite important."

When you lose the commonness of a common species, the consequences are immediate and undeniable. Plainly put, common species are the foundation of ecosystems. They eat or are eaten by other species in large numbers. They influence and engineer their surroundings; in many cases, as with coral reefs and forests, common species effectively *are* the environment.

The most familiar plants and animals are also the ones best known to our eyes, ears, and imaginations, providing touchstones for our relationship with the nonhuman world and even helping to form our sense of place—rock pigeons in New York, Inca doves in Mexico City, peaceful doves in Bangkok. The 20 percent decline in European songbirds (like the 24 common land birds that have declined by 50 to 90 percent in the United States and Canada since 1970) represents a 20 percent reduction in their big-picture roles in pest control, pollination, and seed dispersal, but also in the opportunity to see wild birds and hear birdsong—a hard drop in the small pleasures of being alive.

The role of common species in nature and culture, both physical and metaphysical, all collide in the story of India's white-rumped vultures. They were the subcontinent's primary scavengers of carrion. Without them, putrefying carcasses began to litter the landscape. (India had other vulture species, but each was either far less common to begin with or declined just as steeply as their white-rumped cousins.) The role the vultures played was soon filled by feral dogs, but this led to a spike in fatal dog-bite and rabies cases. The absence of vultures may even have contributed to an outbreak of bubonic plague, when the rotting bodies of cattle killed by a heat wave triggered an explosion of rats that carry the disease.

Until the vulture die-off, traditionalists within the Parsi religious minority had placed their dead in "towers of silence," where the bodies were consumed by the birds, a funeral rite believed to preserve the purity of the Earth; they have now lost that link in what one Parsi leader called the chain of "creation, destruction, and regeneration." Might even the Hindu epic of Ramayana, in which one of the principal figures, Jatayu, is usually depicted as a vulture, lose some quality of meaning in the absence of the actual birds?

None of this, I suspect, is quite convincing enough. I can sense that collective human shrug, the one that says, *we adapt, we survive.* In an India without vultures, people now bury or burn their livestock carcasses, while the Parsi inter their dead in sealed caskets or render them into skeletons using solar-heat concentrators that cost $3,000 apiece. Our common birds are less common than they were, but they are still the most common birds around. Until the recent recovery of the fish stock, the Gulf of Maine had gone decades with very few herring, hardly any whales. We got by, we made do.

We can live without this or that common species; the price is only a little more hardship, a little less awe. It may behoove us to pay attention to the pattern, though. If even the planet's most widespread, abundant, and adaptable species are reaching tipping points, it is because the changes caused by humans have become so overwhelming, inescapable, and complex. "Common species are quite nice bellwethers of the systematic, insidious, and often unrecognized changes and detriments that we're making to the environment," Gaston said.

More and more, the cause of a common species' sudden free fall is impossible to discern, takes years to unravel, or is ultimately given that most modern diagnosis, "multiple causes." They are falling between invisible fault lines, collapsing into the unforeseen fractures of a biosphere under intensifying stress.

Widespread, abundant, adaptable—and damned. Why does it matter what happens to common species? Because we happen to be one of them.

BARACK OBAMA

The Irreversible Momentum
of Clean Energy

FROM *Science*

THE RELEASE OF carbon dioxide (CO_2) and other greenhouse gases (GHGs) due to human activity is increasing global average surface air temperatures, disrupting weather patterns, and acidifying the ocean.[1] Left unchecked, the continued growth of GHG emissions could cause global average temperatures to increase by another 4°C or more by 2100 and by 1.5 to 2 times as much in many midcontinent and far northern locations.[1] Although our understanding of the impacts of climate change is increasingly and disturbingly clear, there is still debate about the proper course for U.S. policy—a debate that is very much on display during the current presidential transition. But putting near-term politics aside, the mounting economic and scientific evidence leaves me confident that trends toward a clean-energy economy that have emerged during my presidency will continue and that the economic opportunity for our country to harness that trend will only grow. This Policy Forum will focus on the four reasons I believe the trend toward clean energy is irreversible.

Economies Grow, Emissions Fall

The United States is showing that GHG mitigation need not conflict with economic growth. Rather, it can boost efficiency, productivity, and innovation.

Since 2008, the United States has experienced the first sustained period of rapid GHG emissions reductions and simultaneous economic growth on record. Specifically, CO_2 emissions from the energy sector fell by 9.5 percent from 2008 to 2015, while the economy grew by more than 10 percent. In this same period, the amount of energy consumed per dollar of real gross domestic product (GDP) fell by almost 11 percent, the amount of CO_2 emitted per unit of energy consumed declined by 8 percent, and CO_2 emitted per dollar of GDP declined by 18 percent.[2]

The importance of this trend cannot be understated. This "decoupling" of energy sector emissions and economic growth should put to rest the argument that combatting climate change requires accepting lower growth or a lower standard of living. In fact, although this decoupling is most pronounced in the United States, evidence that economies can grow while emissions do not is emerging around the world. The International Energy Agency's (IEA's) preliminary estimate of energy-related CO_2 emissions in 2015 reveals that emissions stayed flat compared with the year before, whereas the global economy grew.[3] The IEA noted that "there have been only four periods in the past 40 years in which CO_2 emission levels were flat or fell compared with the previous year, with three of those—the early 1980s, 1992, and 2009—being associated with global economic weakness. By contrast, the recent halt in emissions growth comes in a period of economic growth."

At the same time, evidence is mounting that any economic strategy that ignores carbon pollution will impose tremendous costs to the global economy and will result in fewer jobs and less economic growth over the long term. Estimates of the economic damages from warming of 4°C over preindustrial levels range from 1 percent to 5 percent of global GDP each year by 2100.[4] One of the most frequently cited economic models pins the estimate of annual damages from warming of 4°C at approximately 4 percent of global GDP,[4-6] which could lead to lost U.S. federal revenue of roughly $340 billion to $690 billion annually.[7]

Moreover, these estimates do not include the possibility of GHG increases triggering catastrophic events, such as the accelerated shrinkage of the Greenland and Antarctic ice sheets, drastic changes in ocean currents, or sizable releases of GHGs from previously frozen soils and sediments that rapidly accelerate warming. In addition, these estimates factor in economic damages but do not address the critical question of whether the underlying rate of economic growth (rather than just the level of GDP) is affected by climate change, so these studies could substantially understate the potential damage of climate change on the global macro-economy.[8, 9]

As a result, it is becoming increasingly clear that, regardless of the inherent uncertainties in predicting future climate and weather patterns, the investments needed to reduce emissions—and to increase resilience and preparedness for the changes in climate that can no longer be avoided—will be modest in comparison with the benefits from avoided climate-change damages. This means, in the coming years, states, localities, and businesses will need to continue making these critical investments, in addition to taking commonsense steps to disclose climate risk to taxpayers, homeowners, shareholders, and customers. Global insurance and reinsurance businesses are already taking such steps as their analytical models reveal growing climate risk.

Private-Sector Emissions Reductions

Beyond the macroeconomic case, businesses are coming to the conclusion that reducing emissions is not just good for the environment—it can also boost bottom lines, cut costs for consumers, and deliver returns for shareholders.

Perhaps the most compelling example is energy efficiency. Government has played a role in encouraging this kind of investment and innovation: my administration has put in place (i) fuel economy standards that are net beneficial and are projected to cut more than 8 billion tons of carbon pollution over the lifetime of new vehicles sold between 2012 and 2029,[10] and (ii) 44 appliance standards and new building codes that are projected to cut 2.4 billion tons of carbon pollution and save $550 billion for consumers by 2030.[11]

But ultimately, these investments are being made by firms that decide to cut their energy waste in order to save money and invest in other areas of their businesses. For example, Alcoa has set a goal of reducing its GHG intensity 30 percent by 2020 from its 2005 baseline, and General Motors is working to reduce its energy intensity from facilities by 20 percent from its 2011 baseline over the same timeframe.[12] Investments like these are contributing to what we are seeing take place across the economy: total energy consumption in 2015 was 2.5 percent lower than it was in 2008, whereas the economy was 10 percent larger.[2]

This kind of corporate decision-making can save money, but it also has the potential to create jobs that pay well. A U.S. Department of Energy report released this week found that approximately 2.2 million Americans are currently employed in the design, installation, and manufacture of energy-efficiency products and services. This compares with the roughly 1.1 million Americans who are employed in the production of fossil fuels and their use for electric power generation.[13] Policies that continue to encourage businesses to save money by cutting energy waste could pay a major employment dividend and are based on stronger economic logic than continuing the nearly $5 billion per year in federal fossil fuel subsidies, a market distortion that should be corrected on its own or in the context of corporate tax reform.[14]

Market Forces in the Power Sector

The American electric-power sector—the largest source of GHG emissions in our economy—is being transformed, in large part, because of market dynamics. In 2008, natural gas made up approximately 21 percent of U.S. electricity generation. Today, it makes up roughly 33 percent, an increase due almost entirely to the shift from higher-emitting coal to lower-emitting natural gas, brought about primarily by the increased availability of low-cost gas due to new production techniques.[2, 15] Because the cost of new electricity generation using natural gas is projected to remain low relative to coal, it is unlikely that utilities will change course and choose to build coal-fired power plants, which would be more expensive than natural gas plants, regardless of any near-term changes in federal policy. Although methane emissions from natural gas production are a serious concern, firms have an economic incentive over the long term to put in place waste-reducing measures consistent with standards my administration has put in place, and states will continue making important progress toward addressing this issue, irrespective of near-term federal policy.

Renewable electricity costs also fell dramatically between 2008 and 2015: the cost of electricity fell 41 percent for wind, 54 percent for rooftop solar photovoltaic (PV) installations, and 64 percent for utility-scale PV.[16] According to Bloomberg New Energy Finance, 2015 was a record year for clean-energy investment, with those energy sources attracting twice as much global capital as fossil fuels.[17]

Public policy—ranging from Recovery Act investments to recent tax credit extensions—has played a crucial role, but technology advances and market forces will continue to drive renewable deployment. The levelized cost of electricity from new renewables like wind and solar in some parts of the United States is already lower than that for new coal generation, without counting subsidies for renewables.[2]

That is why American businesses are making the move toward renewable energy sources. Google, for example, announced last month that in 2017, it plans to power 100 percent of its operations using renewable energy—in large part through large-scale, long-term contracts to buy renewable energy directly.[18] Walmart, the nation's largest retailer, has set a goal of getting 100 percent of its energy from renewables in the coming years.[19] And economy-wide, solar and wind firms now employ more than 360,000 Americans, compared with around 160,000 Americans who work in coal electric generation and support.[13]

Beyond market forces, state-level policy will continue to drive clean-energy momentum. States representing 40 percent of the U.S. population are continuing to move ahead with clean-energy plans, and even outside of those states, clean energy is expanding. For example, wind power alone made up 12 percent of Texas's electricity production in 2015, and at certain points in 2015, that number was greater than 40 percent, while wind provided 32 percent of Iowa's total electricity generation in 2015, up from 8 percent in 2008 (a higher fraction than in any other state).[15, 20]

Global Momentum

Outside the United States, countries and their businesses are moving forward, seeking to reap benefits for their countries by being at the front of the clean-energy race. This has not always been the case. A short time ago, many believed that only a small number of advanced economies should be responsible for reducing GHG emissions and contributing to the fight against climate change. But nations agreed in Paris that all countries should put forward increasingly ambitious climate policies and be subject to consistent transparency and accountability requirements. This was a fundamental shift in the diplomatic landscape, which has already yielded substantial dividends. The Paris Agreement entered into force in less than a year, and, at the follow-up meeting this fall in Marrakesh, countries agreed that, with more than 110 countries representing more than 75 percent of global emissions having already joined the Paris Agreement, climate action "momentum is irreversible."[21]

Although substantive action over decades will be required to realize the vision of Paris, analysis of countries' individual contributions suggests that meeting medium-term respective targets and increasing their ambition in the years ahead—coupled with scaled-up investment in clean-energy technologies—could increase the international community's probability of limiting warming to 2°C by as much as 50 percent.[22]

Were the United States to step away from Paris, it would lose its seat at the table to hold other countries to their commitments, demand transparency, and encourage ambition. This does not mean the next administration needs to follow identical domestic policies to my administration's. There are multiple paths and mechanisms by which this country can achieve—efficiently and economically—the targets we embraced in the Paris Agreement. The Paris Agreement itself is based on a nationally determined structure whereby each country sets and updates its own commitments. Regardless of U.S. domestic policies, it would undermine our economic interests to walk away from the opportunity to hold countries representing two-thirds of global emissions—including China, India, Mexico, European Union members, and others—accountable.

This should not be a partisan issue. It is good business and good economics to lead a technological revolution and define market trends. And it is smart planning to set long-term emission-reduction targets and give American companies, entrepreneurs, and investors certainty so they can invest and manufacture the emission-reducing technologies that we can use domestically and export to the rest of the world. That is why hundreds of major companies— including energy-related companies from ExxonMobil and Shell, to DuPont and Rio Tinto, to Berkshire Hathaway Energy, Calpine, and Pacific Gas and Electric Company—have supported the Paris process, and leading investors have committed $1 billion in patient, private capital to support clean-energy breakthroughs that could make even greater climate ambition possible.

Conclusion

We have long known, on the basis of a massive scientific record, that the urgency of acting to mitigate climate change is real and cannot be ignored. In recent years, we have also seen that the economic case for action—and against inaction—is just as clear, the business case for clean energy is growing, and the trend toward a cleaner power sector can be sustained regardless of near-term federal policies.

Despite the policy uncertainty that we face, I remain convinced that no country is better suited to confront the climate challenge and reap the economic benefits of a low-carbon future than the United States, and that continued participation in the Paris process will yield great benefit for the American people, as well as the international community. Prudent U.S. policy over the next several decades would prioritize, among other actions, decarbonizing the U.S. energy system, storing carbon and reducing emissions within U.S. lands, and reducing non-CO_2 emissions.[23]

Of course, one of the great advantages of our system of government is that each president is able to chart his or her own policy course. And president-elect Donald Trump will have the opportunity to do so. The latest science and economics provide a helpful guide for what the future may bring, in many cases independent of near-term policy choices, when it comes to combatting climate change and transitioning to a clean-energy economy.

References and Notes

1. T. F. Stocker *et al.*, in *Climate Change 2013: The Physical Science Basis. Contribution of Working Group I to the Fifth Assessment Report of the Intergovernmental Panel on Climate Change*, T. F. Stocker *et al.*, Eds. (Cambridge Univ. Press, New York, 2013), pp. 33–115.
2. Council of Economic Advisers, in "Economic report of the President" (Council

of Economic Advisers, White House, Washington, DC, 2017), pp. 423–484; http://bit.ly/2ibrgt9.

3. International Energy Agency, "World energy outlook 2016" (International Energy Agency, Paris, 2016).

4. W. Nordhaus, *The Climate Casino: Risk, Uncertainty, and Economics for a Warming World* (Yale Univ. Press, New Haven, CT, 2013).

5. W. Nordhaus, DICE-2016R model (Yale Univ., New Haven, CT, 2016); http://bit.ly/2iJ9OQn.

6. The result for 4°C of warming cited here from DICE-2016R (in which this degree of warming is reached between 2095 and 2100 without further mitigation) is consistent with that reported from the DICE-2013R model in (5), Fig. 22, p. 140.

7. U.S. Office of Management and Budget, *Climate Change: The Fiscal Risks Facing the Federal Government* (OMB, Washington, DC, 2016); http://bit.ly/2ibxJo1.

8. M. Burke, S. M. Hsiang, E. Miguel, *Nature* 527, 235 (2015). doi:10.1038/nature15725 Medline.

9. M. Dell, B. F. Jones, B. A. Olken, *Am. Econ. J. Macroecon.* 4, 66 (2012). doi:10.1257/mac.4.3.66.

10. U.S. Environmental Protection Agency, U.S. Department of Transportation, "Greenhouse gas emissions and fuel efficiency standards for medium- and heavy-duty engines and vehicles—Phase 2: Final rule" (EPA and DOT, Washington, DC. 2016), table 5-40, pp. 5-5–5-42.

11. DOE, Appliance and Equipment Standards Program (Office of Energy Efficiency and Renewable Energy, DOE, 2016); http://bit.ly/2iEHwebsite.

12. The White House, "Fact Sheet: White House announces commitments to the American Business Act on Climate Pledge" (White House, Washington, DC, 2015); http://bit.ly/2iBxWHouse.

13. BW Research Partnership, *U.S. Energy and Employment Report* (DOE, Washington, DC, 2017).

14. U.S. Department of the Treasury, "United States—Progress report on fossil fuel subsidies" (Treasury, Washington, DC, 2014); www.treasury.gov.

15. U.S. Energy Information Administration, "Monthly Energy Review, November 2016" (EIA, Washington, DC, 2015); http://bit.ly/2iQjPbD.

16. DOE, *Revolution . . . Now: The Future Arrives for Five Clean Energy Technologies—2016 Update* (DOE, Washington, DC, 2016); http://bit.ly/2hTv1WG.

17. A. McCrone, Ed., *Clean Energy Investment: By the Numbers—End of Year 2015* (Bloomberg, New York, 2015); http://bloom.bg/2jaz4zG.

18. U. Hölzle, "We're set to reach 100% renewable energy—and it's just the beginning" (Google, 2016); http://bit.ly/2hTEbSR.

19. Walmart, *Walmart's Approach to Renewable Energy* (Walmart, 2014); http://bit.ly/2j5A_PDF.

20. R. Fares, "Texas sets all-time wind energy record" [blog]. *Sci. Am.*, 14 January 2016; http://bit.ly/2iBj9Jq.

21. United Nations Framework Convention on Climate Change, Marrakech Action Proclamation for Our Climate and Sustainable Development (UNFCCC, 2016); http://bit.ly/2iQnUNFCCC.

22. A. A. Fawcett *et al.*, *Science* 350, 1168 (2015). doi:10.1126/science.aad5761 Medline.

23. The White House, *United States Mid-Century Strategy for Deep Decarbonization* (White House, Washington, DC, 2016); http://bit.ly/2hRSWhiteHouse.

Acknowledgments

B. Deese, J. Holdren, S. Murray, and D. Hornung contributed to the researching, drafting, and editing of this article.

DAVID ROBERTS

Wealthier People Produce More Carbon Pollution — Even the "Green" Ones

ONE OF THE perennial debates in environmentalism, which has transferred over to the climate change discussion, is what role personal choices play in the grand scheme of things. Can consumer decisions play a substantial role in reducing emissions? Are people who are concerned about global warming obliged to reduce their own carbon footprint? Is emitting carbon a personal sin (so to speak) as well as a social one?

I have always been a skeptic about the role of personal choices in the climate fight, and a recent study has helped crystallize why. To put it in a nutshell: one's environmental impact is primarily determined by structural features of one's life circumstances, especially socioeconomic status.

Or to put it more bluntly: rich people emit more carbon, even when they recycle and buy canvas tote bags full of organic veggies.

Good Reasons to Reduce Your Footprint

Before I say anything negative about lifestyle decisions (and get a bunch of angry emails), an important caveat: this story focuses only on climate change and carbon emissions.

There are many, many good reasons to live a more environmentally friendly lifestyle. Fresh organic produce is healthy. Reducing/reusing/recycling waste shrinks landfills. Walking and riding bikes makes you happier and more engaged with your community. (Indeed, every second spent locomoting in some fashion other than a personal vehicle is a blessing to your physical and mental health.) Preferring experiences to things is more fulfilling.

Please, go forth and be green. You will be a better person for it.

All I'm talking about here is climate change—what it will take to slow and reverse the rise in global temperature.

Global warming is not only of a different scale than past environmental problems; it is of a different kind. It is not about any one pollutant or set of products. It is deeply and ineradicably systemic, woven into almost everything human beings do here on Earth.

That makes it difficult to connect to, difficult to explain, and extremely difficult to solve. Very big things have to change over very long timescales, and human beings are not generally accustomed to thinking about such things.

That sense of overwhelming scale is part of what motivates the effort to connect climate change to individual behaviors. It gives us something we can wrap our heads around, something that is recognizably within our power.

But the scales of global emissions and personal choices are badly mismatched.

Ecological Footprint Is Mostly Determined by Wealth

The study was published in the June 2017 edition of the journal *Environment and Behavior* with a title that gives you some idea of what to expect: "Good Intents, but Low Impacts."

Spoiler: "Our results show that individuals with high pro-environmental self-identity intend to behave in an ecologically responsible way, but they typically emphasize actions that have relatively small ecological benefits."

The paper, by researchers Stephanie Moser and Silke Kleinhückelkotten (Swiss and German respectively), begins with a review of some past research, which has taken two basic forms.

One line of research, "intent-oriented," has converged on a consensus that what drives environmental behaviors is not any particular set of beliefs about the world, but identity. People who self-identify as "green" do green things. (I suppose this should be no surprise.) That self-identification has been shown to be a better predictor of pro-environment/energy-saving behaviors than other factors like socioeconomic status.

"Impact-oriented" research, however, tells a different story. Study after study finds that the primary determinant of a person's actual ecological footprint is *income*. After that is geography (rural versus urban), various socioeconomic indicators (age, education level, etc.), and household size. Self-identification as "green" is toward the bottom of the list, with mostly marginal effects.

Green Intentions Are Swamped by Wealth

Moser and Kleinhückelkotten set out to test these results by examining detailed data gathered by a survey of about 1,000 representative Germans, done in 2016 for the German Environment Agency.

They found a stark confirmation of previous research. First, "environmental self-identity was the strongest and only significant predictor of pro-environmental behavior."

And second, "environmental self-identity did not predict overall energy use or carbon footprint." In fact, energy use and carbon footprints were slightly *higher* among self-identified greenies. D'oh!

It's not that the pro-environmental behaviors chosen by wealthy, eco-conscious people don't reduce energy use and carbon footprints. They do. Just . . . not very much. And what effect they have is swamped by the much larger effects of wealth, age, and status.

The variables that most predict carbon footprint are "per capita living space, energy used for household appliances, meat consumption, car use, and vacation travel." And wealthy people—even those who self-identify as green—consume more and do more of all those things.

Basically, research shows that the cynical view is roughly correct: Environmental identity will lead to some relatively low-impact (high-signaling) pro-environmental behaviors, but it rarely drives serious reductions in the biggest sources of lifestyle emissions. Environmental self-identification rises with income, but so do emissions.

(A 2012 study and a 2013 study, both based on a survey in Hungary, found roughly the same thing.)

It's Hard to Stray Far from One's Tribe

None of this should come as a surprise. People's lives are heavily path-dependent—we live among people of roughly our socioeconomic status and do roughly what they do, eat what they eat, get around the way they get around.

Consider what the average upper-middle-class American would actually have to do to make a substantial dent in her carbon footprint. Above all, she would have to drastically cut back on travel—virtually never fly and heavily favor walking and biking. She would have to give up meat and live in a small apartment in a dense, transit-served urban area.

Very few people, including those who think of themselves as environmentally conscious, are willing to make lifestyle changes that drastic. Beyond the practicalities, there is a social cost to diverging so sharply from your family and peer groups. It's not easy.

With wealth comes opportunities for consumption. That's why the global wealthy are responsible for such a grossly disproportionate percentage of "lifestyle consumption emissions."

And here's the thing: These are the emissions connected to your personal choices. But just by living in America (or any wealthy, developed country) and enjoying its shared resources and infrastructure, you are responsible for a certain baseline level of its shared emissions as well.

Even if every American could get their lifestyle emissions down somewhere close to that baseline, it still wouldn't be nearly enough to solve climate change. To reduce Americans' per capita baseline to sustainable levels will require decarbonizing power, transportation, and industry, goals over which most individuals have limited control.

And even fairly radical lifestyle changes are meaningless at the level of global emissions unless they are multiplied by many millions. To imagine lifestyle choices making a substantial dent in global warming is to imagine a goodly portion of the world's rich people voluntarily living a lifestyle that is relatively ascetic even by U.S. middle-class standards.

I just find that very difficult to imagine.

Which Is Most Likely to Change, Human Nature or Human Technology?

Author and environmentalist George Monbiot (whose column pointed me to this research in the first place) is pessimistic about our current trajectory, seeing our rampant consumerism on a crash course with ecological reality. He has no faith that green technology can cut the link between our consumption and greenhouse gas emissions (so-called decoupling). He advocates a new political direction he characterizes as "private sufficiency and public luxury."

For my own part, I'm feeling quite gloomy about human nature these days and have less faith than ever in our collective foresight. I don't see the wealthy—and most people reading this likely count among the global wealthy—agreeing to forego the pleasures of a wealthy lifestyle on behalf of future generations, at least not at any real scale, any time soon.

That leaves decoupling, aiming for that exalted future state of a "circular economy," in which all byproducts become feedstocks for something else and there is no waste. And the only way to move in that direction is through policy and technology innovation.

Such techno-dreams may be forlorn, human ingenuity may not move us fast enough, but I have zero faith that the consumer choices of well-meaning greens will move us any faster.

So, again: go forth and be green. You will be happier and healthier. But do not mistake it for a solution to climate change. Only collective action and collective ingenuity can save us.

PART IV

"So Bad, Even the Introverts Are Here"

Profiles

CERIDWEN DOVEY

Dr. Space Junk Unearths the Cultural Landscape of the Cosmos

FROM *The New Yorker*

WHEN NEIL ARMSTRONG stepped onto the moon in 1969, the first images of his momentous leap were received and relayed by antennas at the Honeysuckle Creek Tracking Station, nestled in the foothills of a remote mountain valley in the Australian Alps, south of Canberra. Nothing now remains of the station but its concrete foundations. Nearby lie the ruins of another Australian satellite-tracking station once owned by NASA, called Orroral Valley, which supported the Space Shuttle Columbia missions, among others; it was closed in 1985 and later demolished. Most visitors gaze upon these abandoned sites and dwell on what has been lost. Alice Gorman looks at them and instead sees what might be found. A pioneer in the emerging field of space archeology, she wants to know what the material culture of the Space Age—the artifacts left behind both on Earth and in outer space—can tell us.

Gorman, who is fifty-three, teaches at Flinders University, in Adelaide, and blogs under the name Dr. Space Junk. A few years ago, she took her students to Orroral Valley to do an archeological survey. While kangaroos hopped between the pylons of the Minitrack antenna, the students used a magnetometer to detect intact subterranean cables, which had once connected antennas to processing units. They did a pedestrian-transect survey—noting discoveries along a set route—which revealed hundreds of plastic zip ties hidden in the grass. These had been snipped or burned open, presumably to release the cables as the antennas were dismantled

—and they led Gorman down the obscure scholarly rabbit hole of zip-tie design, back to their American invention in 1958, virtually the same time that the first satellites were launched into space. Zip ties, it turns out, were crucial to bundling wiring safely.

The point of the exercise, Gorman explained to me recently, was to show her students that even an artifact as unremarkable as a zip tie could be an enabler of space exploration. Archeology can be defined as the study of *material* objects, not just old objects, and Gorman believes that considering things like an oxygen tank from Skylab, a glove dropped on the moon, or a defunct satellite still circling our planet as archeological artifacts helps reframe our relationship to space. How might we think differently, she asks, if we considered space as a cultural landscape with richly layered scientific, political, and religious meanings, instead of as an empty vacuum that anybody with heroic ambitions—and a bank balance to match—can venture into with impunity? By considering our heritage in space, Gorman asks us to think carefully about the next steps we take out into the cosmos.

In 2001, Gorman was working as a consultant on Indigenous heritage in central Queensland. After a day documenting a site containing scarred trees once used by Aboriginal people to make bark canoes, she was sitting on the verandah of her cottage, looking up at the stars. She spotted a speeding twinkle and wondered idly whether it could be a satellite. The open country and the wonder-rush of looking at the stars untouched by light pollution brought back memories of her childhood on a farm, and her dreams, back then, of becoming an astrophysicist or an archeologist. On the farm, there were Aboriginal grinding stones used as doorstops, abandoned wells once dug by Chinese workers, a disintegrating homestead made of rammed earth. She was fascinated by an illustrated children's book on archeology sent by a family friend, about Neanderthals and Neolithic Swiss lake villages. But she also loved deciphering the night sky using a star chart in an encyclopedia her father had purchased from a traveling salesman.

At boarding school, Gorman's confidence in her science abilities was knocked; when she got a low mark in physics, she decided that there was no way she could become an astrophysicist. At university, in Melbourne, she studied classical archeology. On that memorable night under the stars in the outback, she realized that she'd finally hit upon a way to integrate her two passions. She threw herself into a self-guided research project on early satellites and began to work her way back into academia. A few years later, already deep into studying space junk, she was thrilled to discover the work of the late scholar William Rathje, who, like most archeologists, believed that one person's trash is another's treasure, and that studying rubbish can challenge received ideas about what has value and why. In 1999, he'd coined the term "exoarchaeology," in an article on the potential for archeological study of orbital space debris (though his main focus was terrestrial trash—he founded the field of "garbology" in 1973 by launching a study of Tucson's landfill).

NASA estimates that there are now more than 500,000 bits of human-made debris the size of a marble or larger in Earth orbit. The visual models are striking: our planet is like a giant beach ball completely surrounded by a dense layer of M&M's (the debris in low Earth orbit), and also by sparser concentric circles of M&M's (the debris at other altitudes, such as geostationary orbit, mostly used for telecommunications). This doesn't even take into account the millions of pieces of debris that are smaller than one centimeter. The Australian researcher Ben Greene says that within 20 years space could be so clogged as to be unusable.

In a recent evening lecture at the Sydney Observatory, Gorman explained that this trash consists of everything from probes, modules, "satellites, rockets, fairings, bolts, flecks of paint, vented fuel, and even human waste." Some of it can, technically, remain intact and in stable orbit for anything from decades to hundreds of years. Despite international efforts to track this debris, nobody has a complete catalogue of exactly what's up there. The Space Age is only six decades old, and already its documentary record is riddled with gaps.

I'd arrived at the observatory, a stately sandstone building overlooking the dark waters of the harbor and the candy-colored city skyline, expecting to be shown into a lecture hall. Instead, the guard directed me downstairs, to a stuffy room in the basement. Gorman was there as the guest speaker at a meeting of the Sydney Space Frontier Society, a chapter of the National Space Society of Australia. The room was packed, which seemed to have taken the organizers by surprise; the elderly secretary of the society fretted that she hadn't put out enough biscuits.

The society's vision is space settlement—"A Spacefaring Civilisa-
tion with people living and working in thriving communities beyond Earth"—and my quick and admittedly prejudiced scan of the
audience suggested its ethos is that of a fringe movement: open to
anybody, amateurs and experts alike, enthusiasts and the slightly
nutty. At the podium, in a black dress, wearing cat's-eye spectacles,
her curly dark hair loose, Gorman was unfazed. She'd warmed up
the room by asking us to consider our "robot avatars" in space: the
1,400 or so active satellites on which our interlinked, technology-
addicted way of life depends. The irony, of course, is that we barely
manage to look up from our phones to walk down the street, let
alone to gaze up at the night sky—which may seem peaceful, Gorman said, but is in fact heavily overpopulated by our own trash.
The room had gone quiet. We were hooked.

One difficulty unique to Gorman's choice of research topic is
the impossibility of doing actual fieldwork in space, at least for
now. She has to be creative in how she uses her tool kit to join the
dots of our history in space in ways that non-specialists can understand. From the swarms of debris in the slide she'd projected
on the observatory's basement wall, Gorman singled out satellites,
telling their stories as if they were old friends. She introduced us
to Vanguard 1, the oldest human-made object in space, and the
first to use solar power. Launched in 1958 by the United States, the
tiny spacecraft was at the time disparagingly called the "grapefruit
satellite" by Khrushchev. It may have been small but its ambitions
were big: Vanguard 1 proved that Earth is not a perfect sphere but
closer to being ever so slightly pear-shaped.

Next, we met the U.S. Navy–launched TRAAC satellite, which
relayed data on the devastating effects of the Starfish Prime high-
altitude nuclear tests, performed in 1962 (the Soviets did their
own tests the same year). The tests created an artificial radiation
belt around the Earth, which eventually disabled a third of all satellites in orbit at the time, TRAAC included. But it's still up there,
with the bizarre distinction of having taken the first poem into
space, etched on its instrument panel. Thomas G. Bergin was commissioned to write the poem, which Gorman described as "slightly
sinister, as it views human spacecraft as weapons against the gods,
who have until now had us at their mercy," though she also admitted that she's always moved by the final line—"And warm with
human love the chill of space." Paying close attention to such ar-

tifacts can demonstrate, sometimes disquietingly, how random the path to the present has been. TRAAC, for example, is a relic of Cold War nuclear-weapons testing in the years before 1967, when the UN's Outer Space Treaty was signed. That agreement obliged all signatories to accept a definition of space as a global commons, safe from territorial claims and to be used only for peaceful purposes.

The stories that Gorman most likes to tell are those that have been obscured by the grand narratives of space exploration. She draws a web of meaning between launch or support sites on Earth —which are often located in supposedly non-spacefaring nations —to show that the nations that are rich or powerful enough to have a stake in space have always depended on less wealthy nations to get there. She has done research at the Woomera Rocket Range, in South Australia, where Aboriginal people were forced from their traditional lands so that Britain and Australia could do nuclear and missile tests nearby, and which was later used by the United States as a spacecraft-monitoring facility. She's spent time at French Guiana's Kourou rocket launch site, one of the busiest spaceports in the world, which was established in 1964 on land expropriated from local people, and is the site of regular protests, most recently in 2016.

Gorman also likes to revive some of the lost tales of missed opportunities for more global cooperation in space. In one article, she tells the story of how NASA, before the Apollo 11 mission, considered sending a United Nations flag to be planted on the moon, rather than an American one, out of respect for the new Outer Space Treaty. In the end, the agency recommended the Stars and Stripes. "Congress backed this decision and altered NASA's appropriations bill to prevent flags of other nations, or international associations, from being placed on the moon on expeditions funded solely by the U.S.A.," she explains.

The American space archeologist Beth Laura O'Leary is one of Gorman's frequent collaborators and another founder of the field. Like Gorman, O'Leary, who is sixty-six, has a background working in cultural-resource management in First Nations communities and had always dreamed of working in space—"but there weren't any female astronauts when I was young," she told me recently. She was excited to discover Gorman's early space-junk research and to connect with a peer thinking along the same lines. At the 2003 World Archeological Congress in Washington, D.C., the two of them cohosted (with John Campbell, an underwater archeologist also interested in space artifacts) the first conference session dedicated to space archeology.

In 2000, O'Leary and one of her past students at New Mexico State University, Ralph Gibson, led a small team to launch the Lunar Legacy Project, an effort to inventory the human-created features and the more than 100 artifacts abandoned at the Apollo 11 landing site—everything from food bags to boots, defecation-collection devices, and even the plastic covering used for the U.S. flag. The astronauts had eight minutes in which to decide what to dump and what to take back to Earth. It struck O'Leary that this was similar to the "drop zone" and "toss zone," which her mentor, Lewis Binford, had observed while doing fieldwork with Nunamiut people in Alaska. In a classic 1978 paper, Binford noted that items entered the archeological record through being dropped (at the hearths where men ate) or tossed (chewed bones chucked over the men's shoulders). O'Leary called up the elderly Binford, who had been on her Ph.D. committee, to tell him about the connection. "That's pretty good," he said. "Eskimos and astronauts." (Gorman pointed out to me that, in both cases, the archeological record also only reflects high-status activities—hunting, visiting the moon—from which women were excluded.)

O'Leary soon realized that the Apollo 11 site fits the criteria for the U.S. National Register of Historic Places, but NASA's lawyers told her that any attempt to frame Tranquility Base as a National Historic Landmark could be perceived as a claim of sovereignty. In 2011, however, O'Leary was invited to provide input on NASA's set of heritage guidelines for anybody planning to go to the moon in the near future. Gorman agrees with O'Leary that the trail of boot prints left by the astronauts in the lunar regolith has enormous significance for our species—as much as the 3.6-million-year-old footprints left by early humans in volcanic ash in Tanzania—yet all it would take to erase them forever is one misguided Google Lunar XPRIZE rover landing in the wrong place.

The controversial U.S. SPACE Act of 2015—the outcome of extensive lobbying by companies set to profit from harvesting space resources—now allows U.S. citizens (and, by extension, corporations) to "engage in the commercial exploration and exploitation" of natural resources found in space, including minerals and water, but excluding biological life. Supporters of the act say that it is a much-needed boost for the commercial space industry. Critics believe that it threatens the Outer Space Treaty's mandate of preventing nations from making sovereign claims in space. Whatever camp you're in, the reality is that commercial entities are going to set foot on the moon, and *soon*. Richard Branson keeps promising that Virgin Galactic's SpaceShipTwo is only a couple of years away from taking the first tourists into space, and Branson's fellow space-cowboy billionaires, Elon Musk and Jeff Bezos, are jostling to launch their own ventures.

Gorman is not opposed to the idea of space tourism per se. But she does think we should see this interpretation of our future in space, often presented as inevitable, as simply one possible future among many. And, in order to choose which future we want, we need to absorb the lessons of the past. Consider the tale of the U.S. Department of Defense's Project West Ford, which launched hundreds of millions of tiny needles, each one a copper dipole antenna, into Earth orbit in the 1960s in order to create an artificial ionosphere to reflect radio signals in case the Soviets cut undersea cables. If the project had continued, it would have killed radio astronomy, since the dipoles blocked radio signals from coming *in* too. One small error for man is always in danger of becoming one giant mistake for humankind.

The morning after Gorman's talk at the Sydney Observatory, I went with her on an informal visit to a research facility in Marsfield, in Sydney's northern suburbs, run by the Australian government's scientific research agency, CSIRO. Lately, Gorman has become interested in the rich histories of large radio-dish antennas, and she knew of two old 13.7-meter antennas on-site, fallen into disuse.

We walked out to a grassy oval beside the office block. Two paraboloid antennas stood on opposite sides of the oval, the peeling paint on their stems still bright enough to give off a glare. Gorman became visibly excited as we approached. "Antennas are monolithic, heavy things, but they get reused and repurposed a lot, because the technology of receiving a signal has essentially stayed the same over time," she said, shielding her eyes as she gazed up at the big dish. We circled the antenna, inspecting the cables still attached to the base. Gorman stopped in her tracks. "Cable ties!" she said, picking up a bit of plastic from the grass.

John Bunton, a renowned CSIRO research scientist with white hair and a white beard, who works on the massive Square Kilometre Array radio telescope, came out to the oval, looking a little suspicious. As soon as Gorman mentioned cable ties, his gruffness evaporated. It turned out that he had been the engineer in charge of the antennas in their original field site in the 1980s, when they'd been part of the highest-resolution radio telescope in the southern hemisphere at the time. When he'd returned to the site years after it was decommissioned, he'd discovered them submerged in water, surrounded by bulrushes, cattle grazing nearby. He pointed to a stain halfway up the antenna's base, the watermark from the flooding. "Do you feel an emotional attachment to these antennas?" Gorman asked. Bunton paused, then smiled. "Well, I call them mine," he said, and for a moment he looked like he might cry.

CAITLIN KUEHN

Of Mothers and Monkeys

FROM *Bellevue Literary Review*

IN THE MORNING I rush past the early shift nurses and the acrid chemical scent of burnt coffee in Styrofoam cups, toward the locked doors near the cafeteria of our city's largest teaching hospital. There are other student research assistants, but I have been here the longest, nearly three years, so I am the most trusted to self-manage the hustle and flow of laboratory life. By the time I layer on the second pair of gray plastic booties over the first set, and steal a look at the clock perched above another layer of locked doors, my mother is already halfway through surgery at the other end of the hospital. I pinch the metal at the bridge of my facemask onto my nose, take a hard breath in, and enter the lab in search of Domingo.

The trick with Domingo is outwitting his speed. He is small enough and jittery enough—and clever enough—that even with the back of the cage slid forward, pressing him with his flank against mesh, he manages to spin himself around until he is spread-eagled facing me. From this position it is impossible to safely administer the ketamine into his quadriceps. Intramuscular injections are a delicate process, a quick and confident plunge through the skin into the muscle tissue before the leg is instinctively jerked away. If the aim is off or I am too self-conscious in my movements, too tenuous, shaky, sluggish, or scared, I risk hitting major arteries or the sciatic nerve that runs deceptively close to the surface in primate thighs. Acute neuropathy, localized necrosis, lameness, and atrophy can all be the permanent fallout of a poorly aimed injection.

Beyond the walls of the lab, and outside the confines of this hospital, it is one of those disconcertingly perfect early October days—all sunshine and light breeze. In here the air is cool and motionless. My supervisors need an early start, so I reach in with my double-gloved hand and tickle gently at the skin between Domingo's leg and pelvis hoping that perhaps he'll flash me a moment of that sacred thigh. Instead he pees.

Domingo is a slender cynomolgus macaque, only about 4.5 kg. The rule of thumb is 0.1 ml of the sedative per kg but I draw up 0.6 ml of ketamine into my syringe. Domingo is an addict baby and so requires more. Before he was born—and before I worked for the lab—Domingo had a mother nicknamed Big Mama because she was particularly large for a female cyno and sometimes researchers are not so creative. Domingo was only discovered when Big Mama was scheduled for euthanasia. She was spared from death by her pregnancy, but Domingo had not been spared from the ketamine he had been exposed to while in utero.

Domingo plopped into the world a wind-up toy. He bounced from one end of the enclosure to another, frantically shaking at the metal doors and the enrichment mirror hanging in the corner. He chewed up his neon-orange toys, tossed them violently across the cage, and smacked and rattled at his mother's head until Big Mama bit his hand off.

This is an image I hold, despite my not being there to witness it—Big Mama in all her imposing girth staring calmly at her caretakers while her tiny anxious child lies on the floor bleeding out. They never did find the hand. She likely ate it. And so, Big Mama was once again scheduled for a large pink syringe of pentobarbital.

This is what came into my mind earlier this morning, just before rushing to work, as I held my own mother's hand while she was prepped for her lumpectomy. She was scared. But I couldn't stay long enough to wait with her and my stepfather until she fell asleep. I had my own anesthetizing to do.

Once my mother begins chemo a few weeks later, it becomes logistically easier for me to rotate between the animal ward and the human ward. When her HCG levels come back slightly positive on the tests prior to her first session, we joke about how—after needing to adopt her two kids—she somehow managed to get pregnant now, at age fifty-three. But still, they delay the first treatment until they are sure.

Her second chemotherapy run, four weeks later at the end of a strangely mild November, does not go well. Sweating and gasping, my mother tells us she can taste tin in her mouth, that her throat is swelling and she can't breathe. My stepfather holds her hand tight as her face goes red in pain and panic. A physician storms in to turn off the chemotherapy cocktail that is flowing too quickly into her body. I so badly want to touch my mother, but I am crammed in the opposite corner of the curtained-off cubicle. I pick at the skin on the top of my hands and make a quiet mantra out of *You're going to be okay* instead.

When she has recovered enough, the drip is restarted, at a much slower rate so that her body can adjust to the onslaught of docetaxel and cyclophosphamide. The therapy is now paired with intravenous Benadryl in hopes that it will stave off the over-zealous defensive mechanisms of her own body. I have to leave, so I tell my stepfather to call, then head to the other end of the hospital.

I'm supposed to be assisting with weekly feline IOPs—a painless tap of the eyeball to measure intraocular pressure, with a fish-flavored treat for good behavior. Even though I am technically allergic to cats, I let Sebastian sit on my lap for a half an hour after the test. But even his rumbling chest against my thigh, the persistent press of his head into my palm, does not calm the palpitations kneading, clawing under my sternum.

Sebastian crawls up the front of my scrub shirt until his face is up against mine. I knock my head against the wall trying to avoid his abrasive kitten tongue. There is no phone service within the cinderblock of the Lab Animal Resources Center, and so I will not receive my stepfather's update until the day's job is complete. There are seven monkeys on the schedule who will need to be sedated for their medical checkups, and then carefully watched during their recoveries, before I'll be able to leave the lab.

As I sit here now, with Sebastian on my lap and a useless phone in my scrub pocket, I wonder what animal first shared with my mother that sudden fear of a throat closing in. But this is probably impossible to know—the history of animal research is too long and labyrinthine to trace. I realize that I—as a student, with very little power but a whole lot of responsibility—am complicit in a moral choice I have still not taken the time to make. Some days it is hard to remind myself that medical research has a purpose. Some days it is as clear as cancer. Some days I just do not know.

Toward the end of December, it has finally snowed, and my mother has her third round of chemotherapy. Today a cat is scheduled for a pattern electro-retinogram. We anesthetize the cat and then flash different-sized stripes of black and white through an ancient computer and record her optical activity. I am responsible for making sure she is breathing appropriately, that her heart rate is okay and blood oxygen levels sufficient. When she is done, I apply a stripe of triple antibiotic ointment to each eye and then softly pinch the skin at the back of her neck. This is the same patch that mother cats gather into their mouths to calm and carry their young. It stretches easily so that I am able to effortlessly inject 60 ml of lactated Ringer's solution underneath her skin from a comically large syringe. The fluid rehydrates her and quickens her recovery from the anesthesia. But even so, she takes longer than most cats to return to the upright position, to return to walking without a drunken slur to her gait.

I wait alone at the hospital cafeteria, checking in on the cat every half hour until 9 p.m. The cafeteria is officially closed, but handfuls of families and late night staff sit at the tables eating ice cream purchased from the kiosk across from the doors to where my cat is attempting to be herself again. My mother is probably at home pecking slowly away at another tray of lasagna left by well-intentioned neighbors.

I stay with my parents for Christmas, which is lovely. The day after is not. I wake up to a Post-it note on the counter informing me that my mother and stepfather rushed to the hospital at 4 a.m. because of a sudden 104° fever. When I call, I am told that my mother has been placed in isolation. The chemotherapy has inadvertently killed off her immune system and now she's fighting off unregulated bacterial growth on top of her own cells' chaotic attempt at manifest destiny. While my mother is trapped alone in a room where people can enter only if dressed in paper suits and those damn plastic booties, I run around the house trying to make things clean and healthy and perfect.

After this, my mother is required to take Neulasta, a medication designed to jump-start the bone marrow's production of white blood cells in order to compensate for the ones she's lost to the chemotherapy. Her doctors hope that taking these injections at home will prevent another sepsis incident. Neulasta is injected subcutaneously, and while my mother is perfectly capable of administering the treatment herself, she is too nervous and cannot manage the syringe. She hands me the refrigerator-chilled box with the premeasured dose and asks me to do it. I have done this 5,000 times, after every cat and monkey procedure. I could do a subcutaneous injection in the dark, and often have, but now I am shatteringly nervous. My mother trusts me, and more than anything I do not want to fuck this up.

Since her diagnosis in October ("Breast Cancer Awareness Month," my mother grumbled at the Packers' pink-outlined shoulder pads. "I am plenty aware."), she has been asking me questions. I mutter what I can about stages, metastasis, faulty cellular junctions, apoptosis, and genetics, but most of it—nearly everything she asks me—I do not know. My medical knowledge is limited to what I have learned in my undergraduate science classes, or here at the lab. All of it applicable only to nonhuman mammals, or else too theoretical to be of any use for as intimate a need as this. I have no good answers. Instead, I breathe in roughly through my nose, tent the skin of her lower abdomen exactly as I have done for the past three years to the soft maternal carrying space of felines and primates, and quick as faked confidence prick the needle in.

I take my time before giving the ketamine to Domingo for today's fundus photography session. The last time I was with him was months ago during my mother's lumpectomy. He is exactly the same. I bring Nutri-Grain fig bars from the monkeys' private kitchen and give some to everyone but Domingo. He will have to wait until after his procedure so as not to upset his stomach, though he tries to snatch his neighbor's treat with his one feisty hand. Domingo requires so much ketamine to sedate and he metabolizes it in strange ways. We'll give him more and more, and there's no effect, no effect, then suddenly he's sound asleep, and then just as suddenly he is wracked with seizures. His hand will clench and his back will arch, his jaw opened like a yawn, or a scream. I'll hold him like an infant, cradled in my arms, as his body tightens in convulsions. "You're going to be okay," I whisper through my mask. "You're going to be okay."

I don't remember when precisely in my mother's treatment she spoke to me of why she was fighting—a term she uses but I, like Susan Sontag, am wary of. The body is not a battlefield, I want to say, and disease and chemotherapy are not soldiers struggling for ownership. We are in my tiny apartment near the hospital. Or maybe we are in the car separated by the console between the front seats. In either case she is wearing one of the three tracksuits she has been alternating for months—today it's the powder blue corduroy.

She adjusts the lay of her stretch cotton cap. "I was trying to think of why I am doing this," she says. She is crying while she speaks. I've swallowed everything into a sock ball stuck in my throat as I always do during these conversations. "It's for you guys," she says. My mother does not cry pretty—no gentle tears, just pink patches of skin and watery eyes. "Your sister and you are why I need to keep fighting. I love you both so much."

It is easily the worst thing she could have said. I almost hate her for it, for mutating this disease into an opportunity for selfless parenting. I want to tell her that mothers aren't meant to be martyrs, that it's . . . it's *selfish*. She's pushed her will to persevere off onto my sister and me. It's too much pressure to be somebody else's reason. But it's impossible to find the words to tell this to someone you love, especially in a moment of mutual vulnerability. It's impossible to ask them to remove you from the calculations. And frankly, it's presumptuous to tell someone how to rationalize their own animal fragility, how to frame their own survival.

In any case, that sock ball has unraveled, that thick ribbed cotton clogging up the drain of my throat. I cough instead. I probably shake my head "no" and hug her. She probably continues to cry into that embrace, with my arms looped awkwardly around her neck, my chest kept carefully two inches away so as not to rub at the sunburn spots of radiation. Or perhaps her hand interlocks painfully tight with mine, balancing on top of the gearshift between us.

By the early spring, my mother is well into her radiation, and I've planned out a tattoo. I too, apparently, want experience at the other end of a needle. It's a Golden Snitch, a childhood symbol of perseverance from my Harry Potter days, with one of those damn ubiquitous ribbons inlaid. Once it's done I'll have to figure out how to bandage the healing skin carefully with tape and gauze before gently tugging down the tight neoprene wrists of my gloves. But today it's just a crudely penned outline that I look at quickly for encouragement. It is going to be a tough day and I don't really want to go in. We have to put down a monkey. I rarely know why.

That's a lie—I do know. Occasionally a lucky cat will be adopted out, fat and happy, to a former student, but this is not the tale for the majority of the cats or monkeys here who have nothing to remedy the ultimate obsolescence of their bodies. For these, euthanasia is written into the end-term of their research protocols. It would be too easy to say that death is a condition of life, but there it is in 12-point font.

What I do in times like these is shave their forearm, to make it easier for my supervisor to see. When I hold the cyno's arm down this morning, his fingers instinctively curl around mine, as my supervisor injects that which will halt his trusting simian heart. My supervisor looks as sad as I am as she waits—watching the cyno's grasp relax from mine, listening for the last faint beat. When she finally pulls the stethoscope from her ears and says it is over, we quietly perfuse the body with a fixative that trickles slowly from two separate IV bags. She soberly cuts off his head with a steel-bladed hacksaw, then settles the body into the double-layered black bags I am holding open. With a more delicate electrical bone saw she carefully chips open his skull to remove the brain, which I place in a jar of buffered 4 percent paraformaldehyde to be donated to another lab in Japan. Death is a simple science.

My first day on the job, three years ago at age nineteen, my supervisor tossed me a live monkey. On the second day, a dead one. They try to get you used to these things early. Distance is a necessary technique in medical research, one with the potential to succeed as often as it fails.

When you're balancing with a garbage bag of deadweight, your sense of space narrowed to a whitewashed room that smells of bodily fluids, paraformaldehyde, and antiseptic, it's hard to remember you're still just a kid. At home though, the image is my mother lying on top of the homemade quilt on her bed. Her tracksuit jacket—velveteen gray this time—is rolled up past her navel, her stomach tipping out of the pants' elastic waistband. She crunches her face and looks away as I inject. It is here I am most desperate to conjure the practiced adulthood I've learned from the lab.

I wasn't aware until that conversation in the car—or in my apartment—that my mother worries she might die. In my empirical brain she is fine. Stage 1, little evidence of metastasis to the lymph nodes. Diagnosis and prognosis are beyond wonderful. I know the treatment is draining and demoralizing, but I have never actually considered her dying. There is no serious threat—it's stage 1, right?—and besides, she is my mother. She, more than anyone in my life, has always been there. Separation is not only impossible, it's unimaginable.

My mother's own father died before she was my age now (colon cancer, likely genetic). She understands and accepts the tender unanswerable realities. I, on the other hand, am desperate to know things that I cannot comprehend. Cats and monkeys keep being put down while I watch and hold their hands, because that is what is written. It is supposed to be my job, as a researcher and a daughter, to confidently understand and explain. Yet, for months, I have been consulting Wikipedia and Google Scholar, hunting for everything I can find on infiltrating ductal carcinoma and Taxotere, certain I can figure it out, until it is so late into the night I might as well wake up again to return to the secret paths of the hospital, the absorbent liners full of monkey pee, the vials of sedative, the aluminum tubes of ointment, the double-layered everything. I'll meet my mother for lunch after her checkup, if she is feeling up to it. Sometimes she comes even if she isn't.

There are days I wish I was raised by Big Mama. My mother would be grumpy and large and not give a shit. Instead of struggling through the conflicting emotions of her own fears and mine, Big Mama would unabashedly claim the space for herself. When I, in my malignant desire to make her body justify itself, or my own self, become too Domingo-like—frantically running back and forth in our enclosure, scratching and fussing at my mother's scalp with its uneven patches of post-chemotherapy soft hair—she would simply bite my hand off. Easy as that. In this story, Big Mama's life is not defined by her maternal moniker, and she is not put down for failing to devote herself to her child. In this story, cancer is not entangled with such sacrificial definitions of motherhood. And in this story, perhaps, Big Mama's anxious bouncing baby would forgive itself all its messy not knowing.

Of course this is a terrible metaphor, one my human mother would ache to hear. I do not truly prefer her as an ornery, old-world primate. What I prefer is the only thing I can properly conceptualize—her *here-ness*. And selfishly, despite all my anger toward her need to center her motivation for survival on me and my sister rather than on herself, what I prefer is a mother who is bright and healthy, who will still place sliced bananas on a plate drizzled with chocolate when I am nervous and unsure. But it does not matter how long your mother saved to purchase you your first microscope, how thinly she sliced the onions for your slides, or how gently she smiled as you settled your head upon her chest and rambled on about topics she didn't understand, mothers—and daughters—are not immortal. I can no longer presume us invulnerable.

PAUL KVINTA

David Haskell Speaks for the Trees

FROM *Outside*

DAVID HASKELL'S BRADFORD PEAR tree stands at the north-west corner of 86th Street and Broadway on Manhattan's Upper West Side, and when we meet there one afternoon in July, he mentions that he hasn't spent quality time with the tree in nearly three months.

The previous occasion he visited, he and his girlfriend, Katie Lehman, were traveling by car to Maine from Sewanee, Tennessee, where Haskell is a professor of biology and environmental studies at the University of the South. They parked on the street, made their way to this corner, and proceeded to loiter, since there's no-where to sit. It was late in the day. Trucks and buses barreled down Broadway. Sirens wailed. Pedestrians flowed past the tree, faces in their phones, while below ground the Seventh Avenue Express hammered by. The tree's fallen white blossoms whirled in the evening gusts, and discarded wads of gum littered the dirt at the base of its trunk. For an hour and a half Haskell watched. He listened. Then he and Lehman got back in their car and drove to Maine.

"It was amazing sharing the tree with Katie, introducing her to this creature I'd spent so much time with," Haskell tells me now. "To be able to wrap other people into my relationship with the tree, and the tree into my relationship with other people—it's very enriching."

Introducing me to the tree, then, is a pretty big deal.

"This is it," he says, beaming.

We eyeball the tree.

"Yes," I say.

It's not exactly beautiful. It's not exactly ugly. It reaches maybe 30 feet tall, with an oval canopy of dark waxy leaves and a gray trunk streaked green with algae. A couple of diseased limbs have been removed, leaving pitted nubs. It grows in front of a Banana Republic, between a newsstand and some newspaper boxes, and nearby there's a flight of stairs leading down to the 86th Street subway platform. At the base of the trunk, some well-tended pink and white periwinkles share a patch of dirt with two cigarette butts, half a grape, a plastic drink lid, and a couple of straws. Locked to the short iron fence that surrounds the trunk is a blue bicycle missing its seat. Another Bradford pear sprouts from the sidewalk 30 feet north of this one, then another one north of that, then another. There are six of them on this block alone.

Haskell's tree is utterly average.

He is not offended by this assessment. In fact, it's one of the reasons he includes the Bradford pear in his book, *The Songs of Trees: Stories from Nature's Great Connectors,* which comes out in April. "This tree appeals to me because it's a regular street tree," he tells me. "There are some trees in Manhattan that are famous, like the 9/11 Survivor Tree. People actually travel great distances to see that tree. No one travels to Manhattan to see this tree." Except Haskell. And now me.

He had invited me to spend a couple of days with him here. I couldn't say no, not after what he had accomplished in his first book, *The Forest Unseen,* a finalist for the 2013 Pulitzer Prize and a book that E. O. Wilson called "a new genre of nature writing, located between science and poetry, in which the invisible appear, the small grow large, and the immense complexity and beauty of life are more clearly revealed." Haskell believes that we live in a world of countless untold stories hiding in plain sight. In *Forest,* he selected a square meter of forest floor and visited that spot almost daily for a year. That's the entire book, all 288 pages of it, him staring at the ground. But Haskell leveraged three remarkable strengths—vast scientific knowledge, prodigious literary gifts, and a deeply meditative approach to fieldwork—to extract from that patch of dirt characters, relationships, drama, and universal themes.

If Haskell could do that in a quiet corner of the forest, I wanted to see what he could come up with on a loud street corner in America's most frenetic metropolis.

Wednesday 7:03 p.m.

An attractive blonde in a short skirt walking three terriers stops under the tree to untangle her leashes. I focus on the woman. Haskell focuses on the dogs. One white puffball refuses to budge when the woman prepares to resume walking. She coaxes the dog. She jerks the leash. "He's saying, 'This is a cool tree,'" Haskell says, meaning literally cool. She's not hearing him. The woman drags the pooch off down the sidewalk.

Haskell strides over to the tree, bends down, and touches the pavement. "Feel that," he tells me. The sidewalk is cool, despite temperatures in the 90s. We then walk out to the median in the middle of Broadway and feel the shade-free pavement there. It's a good 20 degrees hotter than under the tree. "On average, it's 7 degrees warmer in New York City than it is just outside the city, partly because of all these hard surfaces absorbing heat," he says. "But trees change the weather in a city. They have a significant cooling effect. They save a lot on air-conditioning."

7:06 p.m.

Foot traffic is light, probably due to summer vacation. On a normal weekday at this hour, Haskell says, the pedestrian flow would nearly flatten us.

"It's typically a sea of humanity?" I ask.

More like intersecting rivers, he explains. "You've got one coming out of the subway and people flowing north and south. There's a sinkhole with water bubbling up and being drawn back down." There's all this fast water, and then the area around the tree is a quiet pool to the side.

It's illegal to obstruct pedestrian traffic in New York City, Haskell tells me, so if people need to stop they will duck under the tree. That links the plant to the city's sociocultural power dynamics. Haskell calls the area around the tree "gendered and raced space." Over two years, he has seen dozens of folks stop under the tree to check phones or adjust bags. Three-quarters were women of a variety of races; of the men, none was white. Most white guys dominate the middle of the sidewalk, yielding to no one. It's white male privilege, he says, played out on the streets of New York.

7:22 p.m.

Haskell peers into the canopy. "Note the lack of insect damage," he says. A native species would support a riot of caterpillars and leaf miners, munching on leaves, fattening up for predatory birds and spiders. But the Bradford pear hails from China, and Haskell explains that as a foreigner it deploys formidable chemical defenses against local herbivores. This tree ended up here for the same reason Bradford pears ended up across the eastern half of the United States in the 1960s—horticulturalists, smitten by the tree's snowy blossoms, desired an attractive, bug-resistant species for burgeoning suburbs and city beautification projects.

Government officials now classify the tree as a "woody invasive." In 2015, when the Million Trees NYC project realized its goal of planting a million new trees, not one was a Bradford pear. "There was an article spread on Facebook describing them as evil," Haskell says. He's appalled by this. Obviously, native trees are better for the ecological community. But vilifying the Bradford pear denies the full story of our tree here. For starters, it denies what Haskell calls "ancient biogeographical connections," meaning that while this tree is considered a foreigner, it's really not. Millions of years ago, the forests of eastern North America and East Asia were connected, which explains why Bradford pears thrive here. Secondly, human priorities and needs change. "We loved these trees once," Haskell says. "Now we view them as a massive problem. Isn't that more about us and our values than it is about this tree?" What will our needs be in 100 years? Corn, he reminds me, is an exotic species. Due to human need, it has decimated most Midwestern prairies.

Haskell is forty-eight, tall and lanky, with a prominent nose and a bearing that is both slightly formal and slightly awkward. His most distinguishing feature is his accent, which is impossible to place. He was born in England, raised in Paris, and educated at Oxford and Cornell, and he spent the past 20 years in Tennessee. As he has mentioned to journalists before, wherever he goes people tell him: "You're not from here."

The Songs of Trees is similarly global. The book focuses on 12 individual trees around the world. Along with our Bradford pear, the lineup includes a balsam fir in the backwoods of northwestern Ontario, an olive tree at the Damascus Gate in the old city of

Jerusalem, and a giant ceibo deep in the Ecuadorean rainforest, a tree that requires a plane, a bus, two boats, and two days to access. There's a bonsai white pine, two feet tall, that spent its first 350 years in Japan before arriving at the National Arboretum in Washington, D.C., in 1976 as a bicentennial gift. There's a cottonwood sapling in downtown Denver that's been repeatedly reduced to wood chips by beavers. Haskell's hazel tree in Scotland is 10,369 years old. It exists as fingernail-sized bits of black charcoal stored in carefully labeled plastic bags in the Edinburgh offices of a commercial archeology firm.

For two years, Haskell visited all these trees multiple times, spending dozens of hours with each. Day and night, through rain and snow, he watched and listened over long, contemplative stretches. Some sessions were less contemplative than others. One time in Ecuador, having climbed 10 stories up metal ladders attached to the trunk of his ceibo tree, Haskell was taking in the endless biodiversity around him when a bullet ant stabbed him in the neck. "The pain was like a strike on a bell cast from the purest bronze: clear, metallic, single-toned," he writes. Dazed, he flailed at his attacker, only to have it carve a chunk out of his left index finger with its powerful jaws. "Unlike the stinger's purity," he continues in the book, "this pain was a shriek, a fire, a confusion. Over minutes, the sensation ran across the skin of my hand, a cacophony and panic that soaked the hand in sweat. For the next hour my arm was incapacitated." Similarly, he arrived at his olive tree in Jerusalem on one occasion to find its branches "hung with medical equipment and fluorescent safety vests," the gear of Palestinian medics anticipating violence associated with Nakba Day (the "catastrophe" of the founding of Israel). Haskell watched from the tree as security forces slammed into surging protesters, headlocking and dragging several into an armored truck. Still another time, he wandered at night through the dunes of St. Catherines Island off the Georgia coast during a terrifying tropical storm, unable to locate his sabal palm:

> Tonight I discovered that the tree had fallen. Every wave soaks the upturned rootball, and ocean water drowns fronds that, a few days ago, stood atop a nine-meter-tall trunk, lush and vigorous. The fronds were talkative, full of rustle and snap. Now, I hear in them only the detonations and bellow of the sea's quarrel with the land.

Through all this Haskell extracted stories, tales of conflict and cooperation, of life and death. Consider just one example—ants and fungi. High in the crown of the ceibo lives a parasitic fungus, *Ophiocordyceps*, that specializes in invading the body of an ant, consuming it from within, and then somehow commanding it, in its final throes, to anchor itself with its mandibles onto a leaf. From this dangling carcass, infectious fungal spores fall onto the ants beneath. But in other instances, ants and fungi enjoy symbiotic relationships. Below the ceibo, fungi growing inside leafcutter colonies receive a steady supply of fresh leaves and in turn provide meals for the ants. These stories, or "songs" in Haskell's parlance, reveal biological networks—trees networked to other trees, to other plants and animals, to the physical world, to the ancient past. Human beings are very much integrated into these networks, whether the particular tree is located deep in the Amazon or in the heart of Manhattan.

"Muir said that if you want to experience nature, get the hell out of the city!" Haskell tells me, yelling to be heard over a double accordion bus roaring down Broadway. But the very notion of nature stands as a barrier between people and the rest of the community of life, he insists. Cities should be viewed as no more or less natural than a mountain stream running through the so-called wilderness. Noting the urban chaos surrounding us, Haskell says, loudly, "This city is the product of a species that evolved, an advanced primate, *Homo sapiens*." In Haskell's view, Manhattan can't be anything but nature.

We're starting to draw looks. The guy running the newsstand momentarily leaves his post and stares at us. Then he swigs some water from a bottle, spits it out under the tree, and goes back to selling papers.

7:26 p.m.

The Seventh Avenue Express throttles through the subway tunnel two stories beneath us. We feel it under our feet. We also watch it on an app on Haskell's phone, three rippling lines registering the vibration along three different axes. Pressure waves are traveling from the rumbling subway cars into the steel and concrete tunnel, through the ground, and into the iron railing that surrounds the tree and on which Haskell's phone rests. The accelerometer inside his phone captures the movement. Our tree is experiencing the same vibrations as the railing.

In response to decades of train reverberation, the tree has pumped major resources into anchorage, he explains. It has fattened and stiffened its roots with more cellulose and lignin. It hugs the earth tighter than most trees in the forest. Hillside trees do something similar, growing stronger roots along whichever axis the wind typically blows. "This tree is taking the vibratory energy of its environment into its body," Haskell says. The city actually becomes part of the tree. In his book, he explains this by subverting Nietzsche: "What does not kill me becomes part of me, erasing another boundary. Flexure of a tree brings within what was outside. Wood is an embodied conversation between plant life [and] shudder of ground."

7:31 p.m.

A monster dump truck thunders past, grinding its gears. "Did you hear that!" Haskell yells. "Yes!" I yell back. How could I not? "No," he says. "The sparrows." The birds are flitting about the tree's upper branches, swooping down occasionally to fetch crumbs. "I'm hearing the sparrows even though that truck just went by," he says. "If you planted a spectrogram, it would pick up all the low frequencies, like that truck, and the house sparrows would register above that." Sparrows and starlings, he explains, move their calls into higher registers to communicate over the urban rumble. Most bird species can't adapt like that. They lose their acoustic social networks and disappear from urban areas. But sparrows and starlings, along with pigeons, occupy 80 percent of the world's cities. "Their environment has changed them," Haskell says.

7:33 p.m.

Haskell considers our Bradford pear. "That tree isn't an individual," he says. "It's a community." The same could be said for the seemingly autonomous people zipping by—the bike messenger, the woman texting, the guy with the groceries. Just as Bradford pears and house sparrows have incorporated the city into their beings, so too have people, insists Haskell. "We've been yelling and contorting our faces to communicate over the noise," he says. In his book, he cites other examples. "Pitch and genre of

music change our perception of food and wine. A Tchaikovsky waltz . . .
evokes a feeling of sophistication on the tongue that is absent when dining
with a soundtrack of rock music." Or consider any of New York's street
food, he says. It's almost always salty or spicy, otherwise you'd hardly taste
it over the city's noise and smells. What we think of as inner thoughts and
judgments, Haskell says, are very much shaped by external networks. The
same rock band performing on this corner would sound louder performing
at the same volume in a national park, because we expect national parks
to be quieter.

When he was a boy, Haskell would often sit still near the pond
in his backyard and just look at things. "It was my disposition as
a kid," he says. His family moved to Paris from London when
Haskell was three, after his father, a physicist, joined the European
Space Agency. His mother was a biologist. When Haskell was six,
he wrote this story: "Once upon a time there was a golden tadpole
and one day he started to grow his hind legs and then he was getting very excited because he was growing his front legs and then a
few day's after his tail went in and he was a frog." His mother, Jean,
was impressed. "Most people think the tail falls off," she says. "But
his story was absolutely biologically correct."

At the British School of Paris, Haskell fell in love with Shakespeare, Philip Larkin, and many other poets. But the British education system soon demanded specialization, and he spent his last
two years at the school and his time at Oxford immersed exclusively
in biology. He wrote his thesis, "Parasites and the Maintenance of
Sexual Reproduction in Blackberries," under the tutelage of William Hamilton, one of the foremost evolutionary biologists of the
twentieth century. At Oxford, Haskell also learned a fair bit from a
pet rat named Bisquit. Watching the rodent range freely about his
apartment, he observed that rats "are all about social bonds with
others. Bisquit had only humans, but rats in the wild live in complex social networks. What one rat learns gets transmitted through
the network. A rat community is like a scaly tailed, hairy super-
brain, figuring out where and what is safe."

At Cornell, Haskell studied ground-nesting wood warblers. In
his Ph.D. research, he found that the reason chicks don't attract
predators with their cries for food is that high-frequency sounds
travel only short distances in the forest. It was also at Cornell that
Haskell learned to meditate. He described what is now a twice-daily

20-minute practice to me this way: "I sit, and the mental flotsam passes by, sometimes sweeping me into its tangles, sometimes drifting by observed but not entered. I started because I had a sense that my inner disorder needs a practice of trying to pay attention."

After grad school, Haskell took a position at the University of the South, commonly referred to as Sewanee. It was a dream job for an ecologist. Perched atop the Cumberland Plateau, Sewanee encompasses 13,000 acres, 91 percent of it undeveloped forestland. Physically, it's among the largest universities in the country, with the highest diversity of plant species of any campus. Haskell could stroll out of his office and in minutes be in extraordinary old-growth forest.

In time, Haskell became known at Sewanee for his Yoda-like connection to nature. "One day I walked out of the science building, and David mentioned that the tree frogs were peeping," says Marvin Pate, formerly Sewanee's director of sustainability. "It was so subtle. I never would have heard them. If I had, I wouldn't have known what they were." Another time, Haskell was hiking through the forest with a former student, Leighton Reid, who directs restoration-ecology projects around the world for the Missouri Botanical Garden. "He hears something and asks me, 'Are those katydids?'" recalls Reid. "I could barely hear anything. At most it was white noise. And I pay attention to things. I'm in the forest all the time."

Haskell hated the boundaries between academic disciplines and felt scientists needed the arts and humanities. "He's a serious biologist, so there's that scientific side of him," says Jim Peters, a philosophy professor who co-taught "Ecology and Ethics" with Haskell. "But science as purely objective reasoning, he doesn't believe that. Science can help us understand, but it's not pure infallibility. David has an interdisciplinary mind."

Haskell sometimes canceled class so that his students could experience distinguished visitors on campus, of any discipline. They watched Buddhist monks create a mandala. They listened to pianist Jeremy Denk play a concerto. Haskell's "Food and Hunger" course was a multi-subject free-for-all that incorporated two of his passions, meditation and horticulture. (On one acre, Haskell grew most of the vegetables he consumed, along with raising goats, ducks, chickens, rabbits, and bees.) The course explored the ecological aspects of food production, alongside the historical and social aspects of poverty in nearby rural communities. Students practiced a form of *lectio divina*, reading aloud about hunger and then reflecting silently on the text. For Thanksgiving, they prepared a meal for 80 needy local residents.

7:50 p.m.

Haskell shows me some photos. Strolling here today, he had snapped pictures of several tree beds. One shows a trunk surrounded by carefully placed pieces of broken brickwork and creeping ivy. Another has miniature white plastic fencing enclosing what appears to be marijuana growing around the tree. The photos delight Haskell. "These are stories of how people are connected to their trees," he says. The bed beneath our tree is tended by the management of the apartment building on this block, the Belmont. Studies show that the survival rate for trees cared for by neighborhoods in the city is 100 percent, whereas trees that are planted by municipal workers and left on their own have a 60 percent chance of dying within a decade. "Literally, the life of this tree depends on its connection to the community," Haskell says.

It's a two-way relationship. Haskell presses his hand against the trunk and shows me his sooty palm. The tree is filtering the air. Annually, the city's 5 million trees remove 2,000 tons of air pollutants and 40,000 tons of carbon dioxide. New York's tree-planting program now consults maps of asthma hospitalization rates and tree cover in determining which blocks to revegetate.

8:06 p.m.

The howling starts.

A gentleman with wild eyes and terribly mismatched clothes is slouching across Broadway from the other side, coming straight at a group of women who have just exited a yoga class. The racket he's producing contains hints of melodic content, but only hints, like someone singing the blues while getting his prostate checked. For their part, the yogis scatter like billiard balls on the break.

Haskell segues into some ecological play-by-play: "With social networking, you've got all sorts of people manifesting in different ways of being. It's like the interaction between tree roots and fungi. There are a lot of social interactions, but there's also an immune system. If someone seems threatening, you're going to close off. What we're seeing here mirrors what a root is doing when it's conversing with fungi. It's open to conversation. In fact, it will die without conversation, without connection. But if you're open to any kind of connection, you're going to get exploited. A tree root would get overrun with pathogenic fungi and soon die."

Haskell just compared the singing drunk to a deadly fungus.

8:16 p.m.

Haskell's eyes dart skyward. "That high-pitched call," he says. "Kestrel."

I hear nothing. I look up in time to glimpse a black comma soaring high over 86th Street, heading toward Central Park.

Haskell's not an overly emotional guy, but I can tell he's completely jacked up. In two years of observations here, he has spotted exactly five faunal species: house sparrows, starlings, pigeons, one high-flying red-tailed hawk, and one seemingly lost warbler. Kestrels are cavity nesters, so he wonders if someone has erected kestrel boxes in Central Park. "A kestrel is another dimension to the story of this tree, but on a different scale," he says. "It's like connecting a strand from the tree to wherever the bird is headed. It speaks to my excitement of flight. It's flying over the city and seeing the buildings from above."

In 2004, on a cold January morning, Haskell hiked into Sewanee's Shakerag Hollow, wandered off-trail, and stopped only when he found a flat slab of sandstone to sit on. Internally, Haskell had reached a crossroads. He could continue publishing papers with names like "Phylogenetic Analysis of Threatened and Range-Restricted Limestone Specialists in the Land Snail Genus *Anguispira*" that few people read. Or he could try something that accessed more parts of who he is. For some time he had maintained a poetry blog, posting a new haiku every day. And of course he had his meditation practice. What if he combined these three strands —science, meditation, and creative writing? What if he did that right here, in this exact spot in the forest? What might he create?

He had no idea. But it felt right.

Haskell determined to return to this spot over and over. He would come with no agenda, conduct no experiments, collect no specimens. He would simply pay attention. He would later augment his observations with library research. He began calling the meter-square area of ground in front of his rock his "forest mandala," supposing that, just as Buddhist monks believe that the entire universe can be seen through a small circle of colored sand, so too are a forest's ecological stories all present in a mandala-sized area of ground.

What's striking about the essays Haskell subsequently produced aren't necessarily the passages on horsehair worms commandeering the bodies of unsuspecting crickets, or the role of natural selection in shaping our fear of copperheads. That stuff is wonderfully weird and mind-blowing, as is the scene in late January when Haskell almost gets hypothermia after stripping naked at the mandala to compare his body's reaction to the freezing temperature with that of the Carolina chickadee. But the project's real juice flows from his treatment of the least appreciated inhabitants of the mandala — the algae, the fungi, the bacteria. Here's a passage from the book about lichen:

> Lichens don't cling to water as plants and animals do. A lichen body swells on damp days, then puckers as the air dries . . . This approach to life has been independently discovered by others. In the fourth century BCE, the Chinese Taoist philosopher Zhuangzi wrote of an old man tossed in the tumult at the base of a tall waterfall. Terrified onlookers rushed to his aid, but the man emerged unharmed and calm. When asked how he could survive this ordeal, he replied "acquiescence . . . I accommodate myself to the water, not the water to me." Lichens found this wisdom four hundred million years before the Taoists. The true masters of victory through submission in Zhangzi's allegory were the lichens clinging to the rock walls around the waterfall.

Nobody had ever heard of Haskell when Viking published *The Forest Unseen* in 2012, but soon people were comparing him to Henry David Thoreau and Aldo Leopold. The book won a National Academy of Sciences Award, and along with being shortlisted for the Pulitzer Prize, it was runner-up for the PEN/E. O. Wilson Literary Science Writing Award. "I started reading it and thought, 'Oh no, another concept-driven book,'" says Tom Levenson, a Pulitzer judge and professor of science writing at MIT. "The fear is that the author lays out this very clever premise and it won't work. And it's a really constrained premise, one square meter of ground. But he extracts an enormous amount of meaning from that by using incredibly precise poetic language."

Forest was translated into nine languages, including Latvian and two forms of Chinese. Ultimately, the book helped land Haskell a John Simon Guggenheim Fellowship, providing him funding for his next project—listening to trees.

Haskell was interested in arboreal acoustics—wind rustling through branches, raindrops falling on leaves, woodpeckers hammering bark—and what they indicate about ecosystem networks. But he also saw trees as characters that could provide access to the stories of different landscapes across the globe. The overriding theme, as it had been in *Forest,* was connection and relationships, but this time Haskell wanted to explore how humans fit into these networks, both in places where they seemed absent but weren't (the Amazon) and in places where nature seemed absent but wasn't (Manhattan). If people were as connected to the community of life as other organisms, what did that say about the kind of environmental ethic humans should have?

Of course, the idea of listening to a tree is a little weird, especially if you stumble unknowingly upon Haskell doing it. In 2013, Rebecca Hannigan, then a Sewanee sophomore with no knowledge of Haskell's upcoming tree book, attended the school's island-ecology field camp on St. Catherines. Haskell was there to teach but occasionally stole away to visit a particular sabal palm, one of his 12 chosen trees. Late one afternoon, Hannigan spied Haskell alone behind the dunes, holding an audio-recording device beneath the tree. "He was talking into it, then holding it up to the tree, like he was interviewing it and expecting a response," Hannigan recalls. "It was odd."

Thursday 8:40 a.m.

"That guy in the green shirt," Haskell says, "that's Stanley." A 70-year-old African American man is glad-handing his way down the sidewalk. For most of the year except summer, Stanley Bethea sells children's books from under the shade of our tree.

"How ya doin'?" Bethea says, recognizing Haskell. "The tree sure looks good, don't it?"

"It does," says Haskell.

Bethea can't chat long. Kids are clamoring after him. *"They get very upset if I don't speak to them!"* he says.

Had he stayed, Bethea could have told us everything that's blooming in the city right now—the crape myrtles, hydrangeas, hibiscus, everything. *"He's tuned in to the flowering rhythms of this place,"* Haskell says. *"He's been around a long time."*

Ultimately, Haskell contends that guys like Bethea—not academics like himself, or Sierra Club activists, or Washington bureaucrats—are best positioned to make good judgments about landscapes and ecosystems. Bethea is a deeply rooted member of this ecological community, as are the neighborhood folks caring for Manhattan's street trees. They have a mature sense of ecological aesthetics based on belonging, and their ethic will stem from what they view as beautiful and whole. At his olive tree in Jerusalem, Haskell found Bethea's counterparts in Israeli and Palestinian olive farmers. At his ceibo tree in Ecuador, it was the Waorani Indians. *"Embodied, lived experiences within the community of life seems like a pretty good guide to me,"* he says.

8:45 a.m.

A small white butterfly flits by. Haskell is stunned. *"I've never seen a butterfly here,"* he says. It's nothing more than a garden-variety cabbage white, but you'd think he'd just spotted an elusive snow leopard. We're still digesting this historic wildlife sighting when I happen to look up and notice three geese passing overhead.

"No, cormorants," Haskell corrects me. *"Double-crested cormorants! Those are fish-hunting birds. They must be feeding in the rivers."*

Haskell can barely contain himself. There's a direct connection between the city's trees and the Hudson and East Rivers, he explains. Roughly half of New York's sewer system combines sewer and storm runoff, so traditionally, during heavy rains, untreated sewage would back up into the rivers. But trees slow rainwater and divert it into the soil. The city's increased tree cover, combined with sewer improvements, has cleaned up the rivers significantly. There are more fish now, and thus more cormorants.

"In two days we've nearly doubled our species count at the tree," Haskell says, delighted. He stares at the sky in wonder. We watch the cormorants fly toward the Hudson, until they disappear behind tall buildings.

JOSHUA ROTHMAN

A Science of the Soul

FROM *The New Yorker*

FOUR BILLION YEARS ago, Earth was a lifeless place. Nothing struggled, thought, or wanted. Slowly, that changed. Seawater leached chemicals from rocks; near thermal vents, those chemicals jostled and combined. Some hit upon the trick of making copies of themselves that, in turn, made more copies. The replicating chains were caught in oily bubbles, which protected them and made replication easier; eventually, they began to venture out into the open sea. A new level of order had been achieved on Earth. Life had begun.

The tree of life grew, its branches stretching toward complexity. Organisms developed systems, subsystems, and sub-subsystems, layered in ever-deepening regression. They used these systems to anticipate their future and to change it. When they looked within, some found that they had *selves*—constellations of memories, ideas, and purposes that emerged from the systems inside. They experienced being alive and had thoughts about that experience. They developed language and used it to know themselves; they began to ask how they had been made.

This, to a first approximation, is the secular story of our creation. It has no single author; it's been written collaboratively by scientists over the past few centuries. If, however, it could be said to belong to any single person, that person might be Daniel Dennett, a seventy-four-year-old philosopher who teaches at Tufts. In the course of 40 years, and more than a dozen books, Dennett has endeavored to explain how a soulless world could have given rise to a soulful one. His special focus is the creation of the human mind. Into his own he has crammed nearly every related discipline: evolutionary biology, neuroscience, psychology, linguistics, artificial intelligence. His newest book, *From Bacteria to Bach and Back*, tells us, "There is a winding path leading through a jungle of science and philosophy, from the initial bland assumption that we people are physical objects, obeying the laws of physics, to an understanding of our conscious minds."

Dennett has walked that path before. In *Consciousness Explained*, a 1991 bestseller, he described consciousness as something like the product of multiple, layered computer programs running on the hardware of the brain. Many readers felt that he had shown how the brain creates the soul. Others thought that he'd missed the point entirely. To them, the book was like a treatise on music that focused exclusively on the physics of musical instruments. It left untouched the question of how a three-pound lump of neurons could come to possess a point of view, interiority, selfhood, consciousness—qualities that the rest of the material world lacks. These skeptics derided the book as "Consciousness Explained Away." Nowadays, philosophers are divided into two camps. The physicalists believe, with Dennett, that science can explain consciousness in purely material terms. The dualists believe that science can uncover only half of the picture: it can't explain what Nabokov called "the marvel of consciousness—that sudden window swinging open on a sunlit landscape amidst the night of non-being."

Late last year, Dennett found himself among such skeptics at the Edgewater Hotel in Seattle, where the Canadian Institute for Advanced Research had convened a meeting about animal consciousness. The Edgewater was once a rock-and-roll hangout—in the late 1960s and 1970s, members of Led Zeppelin were notorious for their escapades there—but it's now plush and sedate, with overstuffed armchairs and roaring fireplaces. In a fourth-floor meeting room with views of Mount Rainier, dozens of researchers shared speculative work on honeybee brains, mouse minds, octopus intelligence, avian cognition, and the mental faculties of monkeys and human children.

At sunset on the last day of the conference, the experts found themselves circling a familiar puzzle known as the "zombie problem." Suppose that you're a scientist studying octopuses. How would you know whether an octopus is conscious? It interacts with you, responds to its environment, and evidently pursues goals, but a nonconscious robot could also do those things. The problem is that there's no way to observe consciousness directly. From the outside, it's possible to imagine that the octopus is a "zombie"—physically alive but mentally empty—and, in theory, the same could be true of any apparently conscious being. The zombie problem is a conversational vortex among those who study animal minds: the researchers, anticipating the discussion's inexorable transformation into a meditation on *Westworld,* clutched their heads and sighed.

Dennett sat at the seminar table like a king on his throne. Broad-shouldered and imposing, with a fluffy white beard and a round belly, he resembles a cross between Darwin and Santa Claus. He has meaty hands and a sonorous voice. Many young philosophers of mind look like artists (skinny jeans, T-shirts, asymmetrical hair), but Dennett carries a homemade wooden walking stick and dresses like a Maine fisherman, in beat-up boat shoes and a pocketed vest —a costume that gives him an air of unpretentious competence. He regards the zombie problem as a typically philosophical waste of time. The problem presupposes that consciousness is like a light switch: either an animal has a self or it doesn't. But Dennett thinks these things are like evolution, essentially gradualist, without hard borders. The obvious answer to the question of whether animals have selves is that they sort of have them. He loves the phrase "sort of." Picture the brain, he often says, as a collection of subsystems that "sort of" know, think, decide, and feel. These layers build up, incrementally, to the real thing. Animals have fewer mental layers than people—in particular, they lack language, which Dennett believes endows human mental life with its complexity and texture —but this doesn't make them zombies. It just means that they "sort of" have consciousness, as measured by human standards.

Dennett waited until the group talked itself into a muddle, then broke in. He speaks slowly, melodiously, in the confident tones of a man with answers. When he uses philosophical lingo, his voice goes deeper, as if he were distancing himself from it. "The big mistake we're making," he said, "is taking our congenial, shared understanding of what it's like to be us, which we learn from novels and plays and talking to each other, and then applying it back down the animal kingdom. *Wittgenstein*"—he deepened his voice —"famously wrote, 'If a lion could talk, we couldn't understand him.' But no! If a lion could talk, we'd understand him just fine. He just wouldn't help us understand anything about lions."

"Because he wouldn't be a lion," another researcher said.

"Right," Dennett replied. "He would be so different from regular lions that he wouldn't tell us what it's like to be a lion. I think we should just get used to the fact that the human concepts we apply so comfortably in our everyday lives apply only sort of to animals." He concluded, "The notorious *zombie problem* is just a philosopher's fantasy. It's not anything that we have to take seriously."

"Dan, I honestly get stuck on this," a primate psychologist said. "If you say, well, rocks don't have consciousness, I want to agree with you"—but he found it difficult to get an imaginative grip on the idea of a monkey with a "sort of" mind.

If philosophy were a sport, its ball would be human intuition. Philosophers compete to shift our intuitions from one end of the field to the other. Some intuitions, however, resist being shifted. Among these is our conviction that there are only two states of being: awake or asleep, conscious or unconscious, alive or dead, soulful or material. Dennett believes that there is a spectrum, and that we can train ourselves to find the idea of that spectrum intuitive.

"If you think there's a fixed meaning of the word *consciousness,* and we're searching for that, then you're already making a mistake," Dennett said.

"I hear you as skeptical about whether consciousness is useful as a scientific concept," another researcher ventured.

"Yes, yes," Dennett said.

"That's the ur-question," the researcher replied. "Because, if the answer's no, then we should really go home!"

"No, no!" Dennett exclaimed, as the room erupted into laughter. He'd done it again: in attempting to explain consciousness, he'd explained it away.

In the nineteenth century, scientists and philosophers couldn't figure out how nonliving things became living. They thought that living things possessed a mysterious life force. Only over time did they discover that life was the product of diverse physical systems that, together, created something that appeared magical. Dennett believes that the same story will be told about consciousness. He wants to tell it, but he sometimes wonders if others want to hear it.

"The person who tells people how an effect is achieved is often resented, considered a spoilsport, a party-pooper," he wrote, around a decade ago, in a paper called "Explaining the 'Magic' of Consciousness." "If you actually manage to explain consciousness, they say, you will diminish us all, turn us into mere protein robots, mere things." Dennett does not believe that we are "mere *things*." He thinks that we have souls, but he is certain that those souls can be explained by science. If evolution built them, they can be reverse-engineered. "There ain't no magic there," he told me. "Just stage magic."

It's possible to give an account of Dennett's life in which philosophy hardly figures. He is from an old Maine family. By the turn of the eighteenth century, ancestors of his had settled near the border between Maine and New Hampshire, at a spot now marked by Dennett Road. Dennett and his wife, Susan, live in North Andover, Massachusetts, a few minutes' drive from Tufts, where Dennett co-directs the Center for Cognitive Studies. But, in 1970, they bought a 200-acre farm in Blue Hill, about five hours north of Boston. The Dennetts are unusually easygoing and sociable, and they quickly became friends with the couple next door, Basil and Bertha Turner. From Basil, Dennett learned to frame a house, shingle a roof, glaze a window, build a fence, plow a field, fell a tree, butcher a hen, dig for clams, raise pigs, fish for trout, and call a square dance. "One thing about Dan—you don't have to tell him twice," Turner once remarked to a local mechanic. Dennett still cherishes the compliment.

In the course of a few summers, he fixed up the Blue Hill farmhouse himself, installing plumbing and electricity. Then, for many years, he suspended his academic work during the summer in order to devote himself to farming. He tended the orchard, made cider, and used a Prohibition-era still to turn the cider into Calvados. He built a blueberry press, made blueberry wine, and turned it into aquavit. "He loves to hand down word-of-mouth knowledge," Steve Barney, a former student who has become one of the Dennetts' many "honorary children," says. "He taught me how to use a chain saw, how to prune an apple tree, how to fish for mackerel, how to operate a tractor, how to whittle a wooden walking stick from a single piece of wood." Dennett is an avid sailor; in 2003, he bought a boat, trained his students to sail, and raced with them in a regatta. Dennett's son, Peter, has worked for a tree surgeon and a fish biologist, and has been a white-water-rafting guide;

his daughter, Andrea, runs an industrial-plumbing company with her husband.

A few years ago, the Dennetts sold the farm to buy a nearby waterfront home, on Little Deer Isle. On a sunny morning this past December, fresh snow surrounded the house; where the lawn met the water, a Hobie sailboat lay awaiting spring. Dennett entered the sunlit kitchen and, using a special, broad-tined fork, carefully split an English muffin. After eating it with jam, he entered his study, a circular room on the ground floor decorated with sail boat keels of different shapes. A close friend and Little Deer Isle visitor, the philosopher and psychologist Nicholas Humphrey, had emailed a draft of an article for Dennett to review. The two men are similar—Humphrey helped discover blindsight, studied apes with Dian Fossey, and was, for a year, the editor of *Granta*—but they differ on certain points in the philosophy of consciousness. "Until I met Dan," Humphrey told me, "I never had a philosophical hero. Then I discovered that not only was he a better philosopher than me; he was a better singer, a better dancer, a better tennis player, a better pianist. There is nothing he does not do."

Dennett annotated the paper on his computer and then called Humphrey on his cell phone to explain that the paper was so useful because it was so *wrong*. "I see how I can write a reaction that is not so much a rebuttal as a rebuilding on your foundations," he said, mischievously. "Your exploration has helped me see some crucial joints in the skeleton. I hope that doesn't upset you!" He laughed and invited Humphrey and his family to come over later that day.

He then turned to a problem with the house. Something was wrong with the landline; it had no dial tone. The key question was whether the problem lay with the wiring inside the house or with the telephone lines outside. Picking up his walking stick and a small plastic telephone, he went out to explore. Dennett has suffered a heart attack and an aortic dissection; he is robust but walks slowly and is sometimes short of breath. Carefully, he made his way to a little gray service box, pried it open using a multitool, and plugged in the handset. There was no dial tone; the problem was in the outside phone lines. Harrumphing, he glanced upward to locate them: another new joint in the skeleton.

During the course of his career, Dennett has developed a way of looking at the process by which raw matter becomes functional.

Some objects are mere assemblages of atoms to us and have only a physical dimension; when we think of them, he says, we adopt a "physicalist stance"—the stance we inhabit when, using equations, we predict the direction of a tropical storm. When it comes to more sophisticated objects, which have purposes and functions, we typically adopt a "design stance." We say that a leaf's "purpose" is to capture energy from sunlight, and that a nut and bolt are designed to fit together. Finally, there are objects that seem to have beliefs and desires, toward which we take the "intentional stance." If you're playing chess with a chess computer, you don't scrutinize the conductive properties of its circuits or contemplate the inner workings of its operating system (the physicalist and design stances, respectively); you ask how the program is thinking, what it's planning, what it "wants" to do. These different stances capture different levels of reality, and our language reveals which one we've adopted. We say that proteins fold (the physicalist stance), but that eyes see (the design stance). We say that the chess computer "anticipated" our move, that the driverless car "decided" to swerve when the deer leaped into the road.

Later, at a rickety antique table in the living room, Dennett taught me a word game he'd perfected called Frigatebird. Real frigate birds swoop down to steal fish from other birds; in Frigatebird, you steal words made of Scrabble tiles from your opponents. To do so, you use new letters to transform their stems: you can't steal "march" by making "marched," but you can do it by making "charmed." As we played, I tried to attend to the workings of my own mind. How did I know that I could use the letters "u," "t," and "o" to transform Dennett's "drain" into "duration"? I couldn't quite catch myself in the act of figuring it out. To Dennett, this blindness reflects the fact that we take the intentional stance toward ourselves. We experience ourselves at the level of thoughts, decisions, and intentions; the machinery that generates those higher-order properties is obscured. Consciousness is defined as much by what it hides as by what it reveals. Over two evenings, while drinking gin on the rocks with a twist—a "sort of" cocktail—we played perhaps a dozen games of Frigatebird, and I lost every time. Dennett was patient and encouraging ("You're getting the hang of it!"), even as he transformed my "quest" into "equations."

*

A running joke among people who study consciousness is that Dennett himself might be a zombie. ("Only a zombie like Dennett could write a book called 'Consciousness Explained' that doesn't address consciousness at all," the computer scientist Jaron Lanier has written.) The implicit criticism is that Dennett's account of consciousness treats the self like a computer and reflects a disengagement from things like feeling and beauty. Dennett seems wounded by this idea. "There are those wags who insist that I was born with an impoverished mental life," he told me. "That ain't me! I seem to be drinking in life's joys pretty well."

Dennett's full name is Daniel Clement Dennett III. He was born in Boston in 1942. His father, Daniel C. Dennett, Jr., was a professor of Islamic history who, during the Second World War, was recruited by the Office of Strategic Services and became a secret agent. Dennett spent his early childhood in Beirut, where his father posed as a cultural attaché at the American Embassy. In Beirut, he had a pet gazelle named Babar and learned to speak some Arabic. When he was five, his father was killed in an unexplained plane crash while on a mission in Ethiopia. In Dennett's clearest memory of him, they're driving through the desert in a Jeep, looking for a group of Bedouins; when they find the camp, some Bedouin women take the young Dennett aside and pierce his ears. (The scars are still visible.)

After his father's death, Dennett returned to the Boston suburbs with his mother and his two sisters. His mother became a book editor; with some guidance from his father's friends, Dennett became the man of the house. He had his own workshop and, aged six, used scraps of lumber to build a small table and chair for his Winnie-the-Pooh. As he fell asleep, he would listen to his mother play Rachmaninoff's Piano Prelude No. 6 in E-Flat Major. Today, the piece moves him to tears—"I've tried to master it," he says, "but I could never play it as well as she could." For a while, Dennett made money playing jazz piano in bars. He also plays the guitar, the acoustic bass, the recorder, and the accordion, and can still sing the a-cappella tunes he learned, in his twenties, as a member of the Boston Saengerfest Men's Chorus.

As a Harvard undergraduate, Dennett wanted to be an artist. He pursued painting, then switched to sculpture; when he met Susan, he told her that she had nice shoulders and asked if she would model for him. (She declined, but they were married two years later.) A photograph taken in 1963, when Dennett was a graduate student, shows him trim and shirtless in a courtyard in Athens, smoking a pipe as he works a block of marble. Although he succeeded in exhibiting some sculptures in galleries, he decided that he wasn't brilliant enough to make a career in art. Still, he continued to sculpt, throw pots, build furniture, and whittle. His whittlings are finely detailed; most are meant to be handled. A life-sized wooden apple comes apart, in cross-sections, to reveal a detailed stem and core; a fist-sized nut and bolt turn smoothly on minute, perfectly made threads. (Billed as "haptic sculptures," the whittles are currently on display at Underdonk, a gallery in Brooklyn.)

Dennett studied philosophy as an undergraduate with W. V. O. Quine, the Harvard logician. His scientific awakening came later, when he was a graduate student at Oxford. With a few classmates, he found himself debating what happens when your arm falls asleep. The others were discussing the problem in abstract, philosophical terms—"sensation," "perception," and the like—which struck Dennett as odd. Two decades earlier, the philosopher Gilbert Ryle, Dennett's dissertation adviser, had coined the phrase "the ghost in the machine" to mock the theory, associated with René Descartes, that our physical bodies are controlled by immaterial souls. The other students were talking about the ghost; Dennett wanted to study the machine. He began teaching himself neuroscience the next day. Later, with the help of various academic friends and neighbors, Dennett learned about psychology, computer programming, linguistics, and artificial intelligence—the disciplines that came to form cognitive science.

One of Dennett's early collaborators was Douglas Hofstadter, the polymath genius whose book about the mind, *Gödel, Escher, Bach: An Eternal Golden Braid,* became an unlikely bestseller in 1979. "When he was young, he played the philosophy game very strictly," Hofstadter said of Dennett. "He studied the analytic philosophers and the Continental philosophers and wrote pieces that responded to them in the traditional way. But then he started deviating from the standard pathway. He became much more informed by science than many of his colleagues, and he grew very frustrated with the constant, prevalent belief among them in such things as zombies. These things started to annoy him, and he started writing piece after piece to try to destroy the myths that he considered these to be—the religious residues of dualism."

Arguments, Dennett found, rarely shift intuitions; it's through stories that we revise our sense of what's natural. (He calls such stories "intuition pumps.") In 1978, he published a short story called "Where Am I?" in which a philosopher, also named Daniel Dennett, is asked to volunteer for a dangerous mission to disarm an experimental nuclear warhead. The warhead, which is buried beneath Tulsa, Oklahoma, emits a kind of radiation that's safe for the body but lethal to the brain. Government scientists decide on a radical plan: they separate Dennett's brain from his body, using radio transmitters implanted in his skull to allow the brain, which is stored in a vat in Houston, to control the body as it approaches the warhead. "Think of it as a mere stretching of the nerves," the scientists say. "If your brain were just moved over an inch in your skull, that would not alter or impair your mind. We're simply going to make the nerves indefinitely elastic by splicing radio links into them."

After the surgery, Dennett is led into the brain-support lab:

> I peered through the glass. There, floating in what looked like ginger ale, was undeniably a human brain, though it was almost covered with printed circuit chips, plastic tubules, electrodes, and other paraphernalia . . . I thought to myself: "Well, here I am sitting on a folding chair, staring through a piece of plate glass at my own brain . . . But wait," I said to myself, "shouldn't I have thought, 'Here I am, suspended in a bubbling fluid, being stared at by my own eyes'?" . . . I tried and tried to think myself into the vat, but to no avail.

Toward the end of the story, the radio equipment malfunctions, and Dennett's point of view is instantly relocated. It is "an impressive demonstration of the immateriality of the soul, based on physicalist principles and premises," he writes, "for as the last radio signal between Tulsa and Houston died away, had I not changed location from Tulsa to Houston at the speed of light?" The story contains only neurons and machines, and is entirely materialist; even so, it shows that you aren't situated "in" your brain the same way you're situated "in" a room. It also suggests that the intuitions upon which philosophers so confidently rely are actually illusions created by an elaborate system of machinery.

Only rarely do cracks in the illusion of consciousness appear

through which one might see the machinery at work. Proust inspected the state between sleep and wakefulness. Coleridge experi-
mented with mind-altering drugs. Neuroscientists examine minds
compromised by brain injury. Dennett's approach has been to look
back into evolutionary history. In the minds of other animals, even
insects, Dennett believes, we can see the functional components
upon which our selfhood depends. We can also see the qualities
we value most in human selfhood in "sort of" form. Even free will,
he thinks, evolves over evolutionary time. Your amygdala, the part
of the brain that registers fear, may not be free in any meaningful
sense—it's effectively a robot—but it endows the mind to which it
belongs with the ability to avoid danger. In this way, the winding
path leads from determinism to freedom too: "A whole can be *freer*
than its parts."

Along with Richard Dawkins, Sam Harris, and the late Christopher
Hitchens, Dennett is often cited as one of the "four horsemen of
the New Atheism." In a 2006 book called *Breaking the Spell: Religion as a Natural Phenomenon,* he argued that religion ought to be
studied rather than practiced. Recently, with the researcher Linda
LaScola, he published *Caught in the Pulpit: Leaving Belief Behind,*
a book of interviews with clergypeople who have lost their faith.
He can be haughty in his dismissal of religion. A few years ago,
while he was recovering from his aortic dissection, he wrote an
essay called "Thank Goodness," in which he chastised well-wishers
for saying "Thank God." (He urged them, instead, to thank "goodness," as embodied by the doctors, nurses, and scientists who were
"genuinely responsible for the fact that I am alive.")

   Yet Dennett is also comfortable with religion—even, in some
ways, nostalgic for it. Like his wife, he was brought up as a Con-
gregationalist, and although he never believed in God, he enjoyed
going to church. For much of his life, Dennett has sung sacred
music in choirs (he gets misty-eyed when he recalls singing Bach's
"St. Matthew Passion"). He and Susan tried sending their children
to Sunday school, so that they could enjoy the music, sermons, and
Bible stories, but it didn't take. Dennett's sister Cynthia is a min-
ister: "A saintly person," Dennett says, admiringly, "who's a little
annoyed by her little brother."

   The materialist worldview is often associated with despair. In
Anna Karenina, Konstantin Levin, the novel's hero, stares into

the night sky, reflects upon his brief, bubblelike existence in an infinite and indifferent universe, and contemplates suicide. For Dennett, however, materialism is spiritually satisfying. In a 1995 book called *Darwin's Dangerous Idea*, he asks, "How long did it take Johann Sebastian Bach to create the 'St. Matthew Passion'?" Bach, he notes, had to live for 42 years before he could begin writing it, and he drew on 2,000 years of Christianity—indeed, on all of human culture. The subsystems of his mind had been evolving for even longer; creating *Homo sapiens,* Dennett writes, required "billions of years of *irreplaceable* design work"—performed not by God, of course, but by natural selection.

"Darwin's dangerous idea," Dennett writes, is that Bach's music, Christianity, human culture, the human mind, and *Homo sapiens* "all exist as fruits of a single tree, the Tree of Life," which "created itself, not in a miraculous, instantaneous whoosh, but slowly, slowly." He asks, "Is this Tree of Life a God one could worship? Pray to? Fear? Probably not." But, he says, it is "greater than anything any of us will ever conceive of in detail worthy of its detail . . . I could not pray to it, but I can stand in affirmation of its magnificence. This world is sacred."

Almost every December for the past 40 years, the Dennetts have held a black-tie Christmas-caroling party at their home. This year, snow was falling as the guests arrived; the airy modern shingle-style house was decorated like a Yuletide bed-and-breakfast, with toy soldiers on parade. In the kitchen, a small robotic dog-on-wheels named Tati huddled nonfunctionally; the living-room bookshelf displayed a set of Dennett-made Russian dolls—Descartes on the outside, a ghost in the middle, and a robot inside the ghost.

Dennett, dapper in his tuxedo, mingled with the guests. With a bearded, ponytailed postdoc, he considered some mysteries of monkey consciousness; with his silver-haired neighbors, many of whom had attended the party annually since 1976, he discussed the Patriots and the finer points of apple brandy. After a potluck dinner, he called everyone over to the piano, where Mark DeVoto, a retired music professor, was noodling on "O Come, All Ye Faithful." From piles on a Dennett-built coffee table, Dennett and his wife distributed homemade books of Christmas carols.

"Hello!" Dennett said. "Are we ready?" Surrounded by friends, he was grinning from ear to ear. "Let's go. We'll start with 'O Come, All Ye Faithful.' First verse in English, second in Latin!"

Earlier, I'd asked Susan Dennett how their atheism would shape their carol-singing. "When we get to the parts about the Virgin, we sometimes sing with our eyebrows raised," she said. In the event, their performance was unironic. Dennett, a brave soloist, sang beautifully, then apologized for his voice. The most arresting carol was a tune called "O Hearken Ye." Dennett sang the words *"Gloria, gloria / In excelsis Deo"* with great seriousness, his hands at his sides, his eyes faraway. When the carol faded into an appreciative silence, he sighed and said, "Now, that's a beautiful hymn."

Dennett has a philosophical archnemesis: an Australian named David Chalmers. Chalmers, who teaches at NYU and at the Australian National University, believes that Dennett only "sort of" understands consciousness. In his view, Dennett's theories don't adequately explain subjective experience or why there is an inner life in the first place.

Chalmers and Dennett are as different as two philosophers of mind can be. Chalmers wears a black leather jacket over a black T-shirt. He believes in the zombie problem and is the lead singer of a consciousness-themed rock band that performs a song called "The Zombie Blues." ("I act like you act, I do what you do . . . / What consciousness is, I ain't got a clue / I got the Zombie Blues.") In his most important book, *The Conscious Mind,* published in 1996, Chalmers accused Dennett and the physicalists of focusing on the "easy problems" of consciousness—questions about the workings of neurons or other cognitive systems—while ignoring the "hard problem." In a formulation he likes: "How does the water of the brain turn into the wine of consciousness?" Since then, the "hard problem" has been a rallying cry for those philosophers who think that Dennett's view of the mind is incomplete.

Consider your laptop. It's processing information but isn't having experiences. Now, suppose that every year your laptop gets smarter. A few years from now, it may, like IBM's Watson, win *Jeopardy!* Soon afterward, it may have meaningful conversations with you, like the smartphone voiced by Scarlett Johansson in *Her.* Johansson's character is conscious: you can fall in love with her, and she with you. There's a soul in that phone. But how did it get there? How was the inner space of consciousness opened up within the circuits and code? This is the hard problem. Dennett regards it, too, as a philosopher's fantasy. Chalmers thinks that, at present, it is insurmountable. If it's easy for you to imagine a conscious robot, then you probably side with Dennett. If it's easier to imagine a robot that only *seems* conscious, you're probably with Chalmers.

A few years ago, a Russian venture capitalist named Dmitry Volkov organized a showdown between Dennett and Chalmers near Disko Island, off the west coast of Greenland. Before making a fortune investing in Shazam and in the Russian version of PayPal, Volkov was a graduate student in philosophy at Moscow State University, where he wrote a dissertation on Dennett's work. Now he chartered a 168-foot schooner, the S/V *Rembrandt van Rijn,* and invited Dennett, Chalmers, and 18 other philosophers on a weeklong cruise, along with 10 graduate students. Most of the professional philosophers were materialists, like Dennett, but the graduate students were uncommitted. Dennett and Chalmers would compete for their allegiance.

In June, when the Arctic sun never sets, the lowlands of Disko are covered with flowering angelica. The philosophers piled into inflatable boats to explore the fjords and the tundra. The year before, in the *Journal of Consciousness Studies,* Dennett had published a paper called "The Mystery of David Chalmers," in which he proposed seven reasons for Chalmers's resistance to his views, among them a fear of death and a pointless desire to "pursue exhaustively nuanced analyses of our intuitions." This had annoyed Chalmers, but on the cruise the two philosophers were still able to marvel, companionably, at the landscape's alien beauty. Later, everyone gathered in the *Rembrandt*'s spacious galley, where Volkov, a slim, voluble man in sailor's stripes, presided over an intellectual round-robin. Each philosopher gave a talk summarizing another's work; afterward, the philosopher who had been summarized responded and took questions.

Andy Clark, a lean Scottish philosopher with a punk shock of pink hair, summarized Dennett's views. He wore a T-shirt depicting a peacock with a tail made of screwdrivers, wrenches, and other tools. "It obviously looks like something quite colorful and full of complexity and 'peacockness,'" he said. "But, if you look more closely, that complexity is actually built out of a number of little devices."

"A Swiss Army peacock!" Dennett rumbled, approvingly. He was in his element: he loves parties, materialism, and the sea.

After the introduction and summarizing part was over, Chalmers, carrying a can of Palm Belgian ale, walked to the front of the room and began his remarks. Neurobiological explanations of consciousness focus on brain functions, he said. But, "when it comes to explaining consciousness, one needs to explain more than the functions. There are *introspective* data—data about what it's like to be a conscious subject, what it's like experiencing *now* and hearing *now,* what it's like to have an emotion or to hear music." He continued, "There are some people, like Dan Dennett, who think that all we need to explain is the functions . . . Many people find that this is not taking consciousness seriously." Lately, he said, he had been gravitating toward "pan-proto-psychism"—the idea that consciousness might be "a fundamental property of the universe" upon which the brain somehow draws. It was a strange idea, but, then, consciousness *was* strange.

Andy Clark was the first to respond. "You didn't actually give us any positives for pan-psychism," he said. "It was kind of the counsel of despair."

Jesse Prinz, a blue-haired philosopher from CUNY, seemed almost enraged. "Positing dualism leads to no further insights and discoveries!" he said.

Calmly, nursing his beer, Chalmers responded to his critics. He said that he *could* make a positive case for pan-proto-psychism, pointed out that his position wasn't necessarily antimaterialist (a pan-psychic force could be perfectly material, like electromagnetism), and declared that he was all in favor of more neuroscientific research.

Dennett had lurked off to the side, stolid and silent, but he now launched into an argument about perspective. He told Chalmers that there didn't have to be a hard boundary between third-person explanations and first-person experience—between, as it were, the description of the sugar molecule and the taste of sweetness. Why couldn't one see oneself as taking two different stances toward a single phenomenon? It was possible, he said, to be "neutral about the metaphysical status of the data." From the outside, it looks like neurons; from the inside, it feels like consciousness. Problem solved.

Chalmers was unconvinced. Pacing up and down the galley, he insisted that "merely cataloguing the third-person data" could not explain the existence of a first-person point of view.

Dennett sighed and, leaning against the wall, weighed his words. "I don't see why it isn't an embarrassment to your view," he said, "that you can't name a kind of experiment that would get at 'first-personal data,' or 'experiences.' That's all I ask—give me a single example of a scientifically respectable experiment!"

"There are any number of experiments!" Chalmers said, heatedly. When the argument devolved into a debate about different kinds of experimental setups, Dennett said, "I think maybe this session is over, don't you? It's time to go to the bar!" He looked to Chalmers, who smiled.

Among the professional philosophers, Dennett seemed to have won a narrow victory. But a survey conducted at the end of the cruise found that most of the grad students had joined Team Chalmers. Volkov conjectured that for many people, especially those who are new to philosophy, "it's the question of the soul that's driving their opinions. It's the value of human life. It's the question of the special position of humans in the world, in the universe."

Despite his affability, Dennett sometimes expresses a weary frustration with the immovable intuitions of the people he is trying to convince. "You shouldn't trust your intuitions," he told the philosophers on the *Rembrandt*. "Conceivability or inconceivability is a life's work—it's not something where you just screw up your head for a second!" He feels that Darwin's central lesson—that everything in biology is gradual; that it arrives "not in a miraculous, instantaneous whoosh, but slowly, slowly"—is too easily swept aside by our categorical habits of mind. It could be that he is struggling with the nature of language, which imposes a hierarchical clarity upon the world that's powerful but sometimes false. It could also be that he is wrong. For him, the struggle—a Darwinian struggle, at the level of ideas—continues. "I have devoted half a century, my entire academic life, to the project, in a dozen books and hundreds of articles tackling various pieces of the puzzle, without managing to move all that many readers from wary agnosticism to calm conviction," he writes, in *From Bacteria to Bach and Back*. "Undaunted, I am trying once again."

For many years, I took Chalmers's side in this dispute. I read Dennett's *Consciousness Explained*, but I felt that something crucial was missing. I couldn't understand how neurons—even billions of neurons—could generate the experience of being *me*. Terrence Deacon, an anthropologist who writes about consciousness and neuroscience, refers to "the Cartesian wound that separated mind from body at the birth of modern science." For a long time, not even the profoundly informed arguments that Dennett advanced proved capable of healing that wound.

Then, late last year, my mother had a catastrophic stroke. It devastated the left side of her brain, wrecking her parietal and temporal lobes and Broca's area—parts of the brain that are involved in the emotions, the senses, memory, and speech. My mother now appears to be living in an eternal present. She can say only two words, *water* and *time*. She is present in the room—she looks me in the eye—but is capable of only fleeting recognition; she knows only that I am someone she should recognize. She grasps the world, but lightly.

As I spent time with my mother, I found that my intuitions were shifting to Dennett's side of the field. It seems natural to say that she "sort of" thinks, knows, cares, remembers, and understands, and that she is "sort of" conscious. It seems obvious that there is no "light switch" for consciousness: she is present and absent in different ways, depending on which of her subsystems are functioning. I still can't quite picture how neurons create consciousness. But, perhaps because I can take a stance toward my mother that I can't take toward myself, my belief in the "hard problem" has dissolved. On an almost visceral level, I find it easier to accept the reality of the material mind. I have moved from agnosticism to calm conviction.

On a morning this past winter, Dennett sat in an armchair in his Maine living room. The sky and the water were blue and bright. He'd acquired two copies of the *Ellsworth American,* the local newspaper; later, he and Susan would sit by the fireplace and compete to see who could finish the crossword first. In the meantime, he was thinking about the nature of understanding. He recalled a time, many years ago, when he found himself lecturing a group of physicists. He showed them a slide that read "$E=mc^2$" and asked if anyone in the audience understood it. Almost all of the physicists raised their hands, but one man sitting in the front protested. "Most of the people in this room are experimentalists," he said. "They think they understand this equation, but, really, they don't. The only people who really understand it are the theoreticians."

"Understanding, too, comes in degrees," Dennett concluded, back in his Maine living room. "So how do you take that last step? What if the answer is: 'Well, you can only sort of take it'?" Physics, Dennett said, tells us that there are more than three dimensions, and we can use math to prove they're there; at the same time, we struggle to picture them in our heads. That doesn't mean they're not real. Perhaps, he thought, the wholly material soul is similarly hard to imagine. "I'm not ready to say it's unimaginable, because there are times when I think I can imagine it," he said, "and then it doesn't seem to be such a big leap at all. But—it is."

Before the morning slipped away, Dennett decided to go out for a walk, down to where the lawn ended and a rocky beach began. He'd long delighted in a particular rock formation, where a few stones were piled just so, creating a peephole. He was disappointed to find that the tides had rearranged the stones, and that the hole had disappeared. The dock was pulled ashore for the winter, its parts stacked next to his sailboat. He walked down the steps anyway, occasionally leaning on his walking stick. For a few minutes, he stood at the bottom, savoring the frigid air, the lapping water, the dazzling sun.

CHRISTOPHER SOLOMON

The Detective of Northern Oddities

FROM *Outside*

WHEN THEY CAPTURED her off Cohen Island in the summer of 2007, she weighed 58 pounds and was the size of a collie.

The growth rings in a tooth they pulled revealed her age—eight years, a mature female sea otter.

They anesthetized her and placed tags on her flippers. They assigned her a number: LCI013, or 13 for short. They installed a transmitter in her belly and gave her a VHF radio frequency: 165.155 megahertz. Then they released her. The otter was now, in effect, her own small-wattage Alaskan radio station. If you had the right kind of antenna and a receiver, you could launch a skiff into Kachemak Bay, lift the antenna, and hunt the air for the music of her existence: an occasional ping in high C that was both solitary and reassuring amid the static of the wide world.

Otter 13, they soon learned, preferred the sheltered waters on the south side of Kachemak Bay. In Kasitsna Bay and Jakolof Bay, she whelped pups and clutched clams in her strong paws. She chewed off her tags. Some days, if you stood on the sand in Homer, you could glimpse her just beyond Bishop's Beach, her head as slick as a greaser's ducktail, wrapped in the bull kelp with other females and their pups.

"They're so cute, aren't they?" said the woman in the gold-rimmed eyeglasses. She was leaning over 13 as she said this, measuring a right forepaw with a small ruler. The otter's paw was raised to her head as if in greeting, or perhaps surrender. "They're one of the few animals that are cute even when they're dead."

Two weeks earlier, salmon set-netters had found the otter on the beach on the far side of Barbara Point. The dying creature was too weak to remove a stone lodged in her jaws. Local officials gathered her up, and a quick look inside revealed the transmitter: 13 was a wild animal with a history. This made her rare. She was placed on a fast ferry and then put in cold storage to await the attention of veterinary pathologist Kathy Burek, who now paused over her with a sympathetic voice and a scalpel of the size usually seen in human morgues.

Burek worked with short, sure draws of the knife. The otter opened. "Wow, that's pretty interesting," Burek said. "Very marked edema over the right tarsus. But I don't see any fractures." The room filled with the smell of low tide on a hot day, of past-expiration sirloin. A visiting observer wobbled in his rubber clamming boots. "The only shame is if you pass out where we can't find you," Burek said without looking up. She continued her exploration. "This animal has such dense fur. You can really miss something." She made several confident strokes until the pelt came away in her hands, as if she were a host gently helping a dinner guest out of her coat. The only fur left on 13 was a small pair of mittens and the cap on her head, resembling a Russian trooper's flap-eared *ushanka.*

It had been nearly a year since Burek's inbox pinged with notice of a different dead sea otter. Then her email sounded again, and again after that. In 2015, 304 otters would be found dead or dying, mostly around Homer and Kachemak Bay, on Alaska's Kenai Peninsula. The number was nearly five times higher than in recent years. On one day alone, four otters arrived for necropsy. Burek had to drag an extra table into her lab so that she and a colleague could keep pace—slicing open furry dead animals, two at a time, for hours on end.

As they worked, an enormous patch of unusually warm water sat stubbornly in the eastern North Pacific. The patch was so persistent that scientists christened it the Blob. Researchers caught sunfish off Icy Point. An unprecedented toxic algal bloom, fueled by the Blob, reached from Southern California to Alaska. Whales had begun to die in worrisome numbers off the coast of Alaska and British Columbia—45 whales that year in the western Gulf of Alaska alone, mostly humpbacks and fins. Federal officials had labeled this, with an abstruseness that would please Don DeLillo, an Unusual Mortality Event. By winter, dead murres lay thick on beaches. The Blob would eventually dissipate, but scientists feared that the warming and its effects were a glimpse into the future under climate change.

What, if anything, did all this have to do with the death of 13? Burek wasn't sure yet. When sea otters first began perishing in large numbers around Homer several years ago, she identified a culprit: a strain of *streptococcus* bacteria that was also an emerging pathogen affecting humans. But lately things hadn't been quite so simple. While the infection again killed otters during the Blob's appearance, Burek found other problems as well. Many of the otters that died of strep also had low levels of toxins from the Blob's massive algal bloom, a clue that the animals possibly had even more of the quick-moving poison in their systems before researchers got to them. They must be somehow interacting. Perhaps several problems now were gang-tackling the animals, each landing its own enervating blow.

Burek's lilac surgical gloves grew red. She noted that the otter had a lung that looked "weird." She measured a raspberry-sized clot on a heart valve using a piece of dental floss. She started working in the abdomen.

"Huh," she said. She'd noticed that the lower part of the otter was filled with brown matter and bits of shell: the nearly digested remains of the animal's last meals had spilled into its pelvis and down into a leg wound. This could have caused an infection and also led to blood poisoning. But where was the injury?

"The colon got perforated. I have no idea how," Burek said. She probed further until she found a pocket of something like pus at the top of the femur. She eventually separated the femur from the body, and her assistant placed the bone in a Ziploc bag.

By now it was past lunchtime. Burek had been at the necropsy table for more than three hours without pause. She looked a little weary. What caused the otter's death would remain, for the moment, unresolved. The not knowing seemed to displease her, though Burek was accustomed to mystery. The frozen north was always shifting; you took it as you found it.

Burek straightened stiffly. "I'm hungry," she said across the bloody table. She removed the otter's head and reached for the bone saw. "Who likes Indian?"

*

Burek often spends her days cutting up the wildest, largest, small-est, most charismatic, and most ferocious creatures in Alaska, look-ing for what killed them. She's been on the job for more than 20 years, self-employed and working with just about every organiza-tion that oversees wildlife in Alaska. Until recently, she was the only board-certified anatomic pathologist in a state that's more than twice the size of Texas. (There's now one other, at the Uni-versity of Alaska.) She's still the only one who regularly heads into the field with her flensing knives and vials, harvesting samples that she'll later squint at under a microscope.

Nowhere in North America is this work more important than in the wilds of Alaska. The year 2015 was the planet's hottest on record; 2016 is expected to have been hotter still. As human-generated greenhouse gases continue to trap heat in the world's oceans, air, and ice at the rate of four Hiroshima bomb explosions every second, and carbon dioxide reaches its greatest atmospheric concentration in 800,000 years, the highest latitudes are warming twice as fast as the rest of the globe. Alaska was so warm last winter that organizers of the Iditarod had to haul in snow from Fairbanks, 360 miles to the north, for the traditional start in Anchorage. The waters of the high Arctic may be nearly free of summertime ice in little more than two decades, something human eyes have never seen.

If Americans think about the defrosting northern icebox, they picture dog-paddling polar bears. This obscures much big-ger changes at work. A great unraveling is under way as nature gropes for a new equilibrium. Some species are finding that their traditional homes are disappearing, even while the north becomes more hospitable to new arrivals. On both sides of the Brooks Range—the spine of peaks that runs 600 miles east to west across northern Alaska—the land is greening but also browning as tun-dra becomes shrubland and trees die off. With these shifts in cli-mate and vegetation, birds, rodents, and other animals are on the march. Parasites and pathogens are hitching rides with these new-comers. "The old saying was that our cold kept away the riffraff," one scientist told me. "That's not so true anymore."

During this epic reshuffle, strange events are the new normal. In Alaska's Arctic in summertime, tens of thousands of walruses haul out on shore, their usual ice floes gone. North of Canada, where the fabled Northwest Passage now melts out every year, satellite-tagged bowhead whales from the Atlantic and Pacific recently met for the first time since the start of the Holocene.

These changes are openings for contagion. "Any time you get an introduction of a new species to a new area, we always think of disease," Burek told me. "Is there going to be new disease that comes because there's new species there?"

A lot of research worldwide has focused on how climate change will increase disease transmission in tropical and even temperate climates, as with dengue fever in the American South. Far less attention has been paid to what will happen—indeed, is already happening—in the world's highest latitudes, and to the people who live there.

Put another way: the north isn't just warming. It has a fever.

This matters to you and me even if we live thousands of miles away, because what happens in the north won't stay there. Birds migrate. Disease spreads. The changes in Alaska are harbingers for what humans and animals may see elsewhere. It's the frontline in climate change's transformation of the planet.

This is where Burek comes in. Fundamentally, a veterinary pathologist is a detective. Burek's city streets are the tissues of wild animals, her crime scenes the discolored and distended organs of tide-washed seals and emaciated wood bison. "She's the one who's going to see changes," says Kathi Lefebvre, a lead research biologist at Seattle's Northwest Fisheries Science Center, a division of the National Oceanic and Atmospheric Administration (NOAA). "She's the one who's going to see epidemics come along. And she's the one with the skills to diagnose things."

As the planet enters new waters, Burek's work has made her one of the lonely few at the bow, calling out the oddness she sees in the hope that we can dodge some of the melting icebergs in our path.

It's a career that long ago ceased to strike Burek as unusual, and she moves without flinching through a world tinged with blood and irony. The first time we spoke on the phone, Burek offhandedly said of herself and a colleague, "We've probably cut up more sea otters than anybody else on the planet."

"Congratulations," I said.

"We all got to brag about something," she replied.

Summer is the season when Alaskans at play under the undying sun tend to come across dead or stranded animals and place a call to a wildlife hotline. The call starts a chain of events that often ends at Burek's family home, which is made of honey-colored logs and sits on an acre and a half in Eagle River, about 20 miles north of downtown Anchorage. This is where the asphalt yields to Alaska. The rough peaks of the Chugach Mountains, still piebald with snow in midsummer, lean overhead. Moose occasionally carry off the backyard badminton net in their antlers.

In July, I headed north from Seattle to spend a month with Burek as she worked. She's fifty-four but looks a decade younger, with long brown hair and appled cheeks that give her the appearance of having just come in from the icicled outdoors. Her voice has an approachable Great Plains flatness, the vestige of her Wisconsin birth and an upbringing in the suburbs of Ohio. Burek ends many sentences with a short, sharp laugh—a punctuative caboose that can signal either amusement or bemusement, depending. Growing up in the Midwest, she didn't see the ocean until high school. "But I was always fascinated by whales," she told me. "And I always wanted to be a vet or a wildlife biologist—Jane Goodall or something." She laughed. "Lots of kids wanted to be vets. They outgrow it."

Burek was intrigued by the biology—how bodies worked and how, sometimes, they didn't. After college she went to veterinary school at the University of Wisconsin–Madison, later moving to Alaska to see how she would like working in a typical vet practice. One year she lived outside Soldotna, in a one-room "dry" cabin, with no running water, while writing her thesis for a master's degree in wildlife disease virology. Alaska agreed with her. "I like the seasons. I like the wilderness. I like the animals," she said.

Burek met her future husband, Henry Huntington, on the coast of the Chukchi Sea in the high Arctic, during the Inupiat's annual spring bowhead whale hunt, when breezes pushed the ice pack together and forced a pause in the whaling. They now have two teenage sons. "I tell the boys they're the product of persistent west winds in May of 1992," said Henry, a respected researcher and scientist with the Pew Charitable Trusts' Arctic conservation campaigns.

Surprisingly little is known about the diseases of wildlife. As a result, many veterinary pathologists end up focusing on a few species. Thanks to Burek's curiosity and her gifts, and to a necessary embrace of the Alaskan virtue of do-it-yourself, her expertise is broad. "Anyone who gets into this kind of thing, you like a puzzle," she told me. "You have to pull together all kinds of little pieces of information to try to figure it out, and it's very, very challenging."

Over the years, Burek has peered inside just about every mammal that shows up in Alaskan field guides. One morning, as we drank coffee at her kitchen table, she rattled off a few dozen examples. Coyotes. Polar bears. Dall sheep. Five species of seals. As many whales, including rare Stejneger's beaked whales.

As we talked, I wandered into the living room. On a wall not far from the wedding photos hung feathery baleen from the mouths of bowhead whales and the white scimitars of walrus tusks. Upstairs in a loft lay an oosik—the baculum, or penis bone, of another walrus. It was as long as a basketball player's tibia. Atop the fireplace mantel, where other families might display pictures of wattled grandparents, grinned a row of skulls: brown bear, lynx, wood bison. Burek tapped one of the skulls in a spot that looked honeycombed. "Abscessed tooth," she said. "Wolf. One of my cases."

Working on wild animals, often in situ, routinely presents her with job hazards that simply aren't found in the Lower 48. Anchorage sits at the confluence of two long inlets. When Burek performs necropsies on whales on Turnagain Arm, she has to keep a sentry's eye on the horizon for its infamous bore tide, when tidal flow comes in as a standing wave, fast enough that it has outrun a galloping moose. Knik Arm is underlain in places by a fine glacial silt that, when wet, liquefies into a lethal quicksand. Burek's rule of thumb in the field is never to sink below her ankles. Not long ago, while taking samples from a deceased beluga, she kept slipping deeper. Exasperated, she finally climbed inside the whale and resumed cutting.

Then there's the problem of the whales themselves. "Whales are just like Crock-Pots," Burek said. "They're kind of encased in this thick layer of blubber that's designed to keep them warm. They might look okay on the outside, but inside everything is mush."

Decay is the nemesis of the pathologist. Decay erodes evidence. "Fresher is always better," Burek said, sounding like a discerning sushi chef. It isn't possible every time. Colleagues told me about a trip with Burek to a remote beach outside Yakutat, to do a postmortem on a humpback. There were several in the group, including a government man with a shotgun to keep away the brown bears that sometimes try to dine on Burek's specimens. It was raining and cold, and the whale had been dead for a while. Inside, the organs were soup. The pilot who retrieved them had to wear a respirator.

"My wife," Henry told me, "has a high threshold for discomfort."

One morning in Anchorage, my phone buzzed. To get a text from Burek is to gain new appreciation for the cliché *mixed emotions*. Often it's a chirpy message notifying you that another of God's creatures has expired and would you like to come see the carcass?

Burek picked me up at a coffee shop on Northern Lights Avenue, driving the family's Dodge Grand Caravan with a cracked windshield. Outside it was sunny and warm; just two days earlier, it hit 85 in Deadhorse, the highest temperature ever recorded on the North Slope. Burek's eyeglasses were covered by sun blockers of the type sold on late-night television. She was wearing summer sandals, her toenails painted what a saleswoman would call "aubergine." Her foot pressed the gas. We were going to pick up a dead baby moose.

"Fish and Game wants to know why it died, if it's a possible management issue," Burek said. Last year an adenovirus, which is more commonly seen in deer in California, had killed two moose in Alaska. Officials wanted to know how common adenovirus was in the state.

As work went, it was an unremarkable day for Burek. The past several years had presented her with a string of cases that were altogether more intriguing and odder and more frustrating for their open-endedness. In 2012, Burek and others observed polar bears that had suffered a curious alopecia, or hair loss, but they were unable to pinpoint the reason. In 2014, Burek described a sea otter that had died of histoplasmosis, an infection caused by a fungus that's usually found in the droppings of Midwestern bats. The infection will sometimes afflict spelunkers, which is where it gets its common name: cave disease. The finding was a dubious first, both for a marine mammal and for Alaska. But, again, why? What was a Midwestern fungus doing inside an otter plashing off the coast of Alaska?

Then there was the strange case of the ringed seals. In the spring of 2011, Native hunters in Barrow, the northernmost town in the United States, started finding ringed seals that didn't look right. The animals had lesions around their mouths and eyes, and ulcers along their flippers. Some had gone bald. A handful died.

Soon, down the coast at the major walrus rookery at Point Lay, ulcers started turning up on walruses both living and dead. The number of sightings on spotted and bearded seals increased and spread south into the Bering Strait as the summer progressed. In time, a few ribbon seals were also affected. Federal officials labeled it another Unusual Mortality Event, a signal of concern and a call for more study. Burek led the postmortems, opening up dozens of animals. Researchers sent samples as far away as Columbia University, in New York, for molecular work. They tested many ideas, but the cause eluded them.

Was climate change a factor in the events? The evidence intrigued Burek and her colleagues. Seals molt during a brief span of time in the spring. According to Peter Boveng, the polar ecosystems program leader at NOAA's Alaska Fisheries Science Center, the longer days and warming temperatures likely cue the animals to climb onto the sea ice, so their skin can warm up and start the process of dropping old hairs and growing new ones. Having ice present is probably crucial for this molting process to happen, Boveng and others believe.

But what if a warming north meant less ice for the seals to use, interfering with their molt? That would explain why the animals showed lesions in the same places on their bodies where the molt begins—the face, the rear end. And when the skin is unprotected by fur, Burek told me, "it may be susceptible to secondary infections" from bacteria and fungi in the environment.

Nature, alas, is messy and confusing. Though the reasoning seemed plausible, there was no widespread lack of spring ice in 2011 in the areas where the diseased seals were found. Deepening the mystery, lesions in walruses all but vanished in subsequent years, even as some seals continue to have them. "It's very frustrating—very frustrating," Burek said of trying to tease out an answer. A lot of her work remains unresolved. Burek knows that this is the reality of doing her job in the 49th state. It is a vast place, expensive to do research; scientists often haven't been able to do enough baseline studies to know what's normal and expected, versus new and worrisome, in a given population. Still, it chews at her, the inability to give answers to concerned Native peoples. "I have enough self-doubt that it's like, well, maybe it's because I'm not working hard enough, or I haven't done the right thing to figure it out."

To be sure, the far north isn't collapsing under contagion caused by climate change. And Burek is careful about drawing connections. Still, a good detective doesn't need a smoking gun to know when a crime has been committed. Circumstantial evidence, if there's enough of it, and the right kinds, can tell the story. "It seems hard to believe," Burek told me, "that a lot of these changes aren't related to what's going on in the environment. The problem is proving it."

There's a larger question, too, about what these developments augur for humans. The answer, researchers are finding, is that it's already starting to matter.

Time was, the cold and remoteness of the far north kept its freezer door closed to a lot of contagion. Now the north is neither so cold nor so remote. About 4 million people live in the circumpolar north, sometimes in sizable cities (Murmansk and Norilsk, Russia; Tromsø, Norway). Oil rigs drill. Tourist ships cruise the Northwest Passage. And as new animals and pathogens arrive and thrive in the warmer, more crowded north, some human sickness is on the rise too. Sweden saw a record number of tick-borne encephalitis cases in 2011, and again in 2012, as roe deer expanded their range northward with ticks in tow. Researchers think the virus the ticks carry may increase its concentrations in warmer weather. The bacterium *Francisella tularensis,* which at its worst is so lethal that both the United States and the USSR weaponized it during the Cold War, is also on the increase in Sweden. Spread by mosquitoes there, the milder form can cause months of flu-like symptoms. Last summer in Russia's far north, anthrax reportedly killed a grandmother and a boy after melting permafrost released spores from epidemic-killed deer that had been buried for decades in the once-frozen ground.

Alaska hasn't been immune to such changes. A few months ago, researchers reported that five species of nonnative ticks, probably aided by climate change, may now be established in the state. One is the American dog tick, which can transmit the bacterium that causes Rocky Mountain spotted fever, which can lead to paralysis in both canines and humans. In 2004, a bad case of food poisoning sent dozens of cruise-ship passengers running to their cabins.

The culprit was *Vibrio parahaemolyticus,* a leading source of seafood-related food poisoning. *V. parahaemolyticus* is typically tied to eating raw oysters taken from the warm waters of places like Louisiana. Why was it infecting people 600 miles north of the most northerly recorded incident? Health officials later teased out the reason: summer water temperatures in Prince William Sound, where the oysters are farmed, now gets warm enough to activate the bacterium.

Earlier in 2016, Burek and NOAA's Lefebvre coauthored a paper about their discovery of domoic acid in all 13 species of Alaskan marine mammals they examined, from Steller sea lions to humpback whales, in waters as far north as the Arctic Ocean. Domoic acid is naturally produced by some species of algae, and it moves through the food web as it accumulates in the filter-feeding animals that dine on it—anchovies, sardines, crabs, clams, and oysters. Scientists knew the algae that makes domoic acid were present, but they never had a report of a bloom that far north before 2015. The hunch is that warming waters may be increasing the toxin's presence in Alaska.

"What's going to happen to these 100-year-old whales when they get hit by these neurotoxins three years in a row?" Lefebvre said. "And it's not just mortality. It's sub-lethal neurological effects."

A study published in 2015 in the journal *Science* found that harmful algal blooms off the California coast have caused enough brain damage to California sea lions that they lose their way and have trouble hunting. "This is a shot across the bow," Lefebvre said of the algal blooms. "It's the type of thing that could happen and become more common."

Here's the broader lesson: if the animals can get sick, we can get sick, whether it's from invigorated pathogens in the environment or from ailing animals themselves. Three in four emerging infectious diseases in humans today are zoonotic diseases—illnesses passed from animals to humans.

This is one reason Burek has a soft spot for sea otters like 13: they are excellent sentinels for what's happening in the world. Otters splash in the same waters where humans live, work, and play. They eat the same seafood humans do. "I call them a pathologist's wonderland, because they get all the fantastic, extreme infectious diseases—not to sound too unpleasant," Burek said.

There are other reasons to pay attention to animals like otters.

Mike Brubaker, director of community environment and health at the Alaska Native Tribal Health Consortium, points out that traditional foods—everything from salmonberries to moose meat—still make up 80 percent of the Native diet in some remote Alaskan communities. If animals suffer, then traditional diets suffer, and so do the cultures that revolve around hunting, fishing, and foraging.

Making Burek's job even more complicated, animals frequently die from mysterious causes that may have nothing at all to do with climate change. As she pokes through the bones, her constant challenge is to discern what's notably weird from what's simply everyday and unfortunate.

Near the airport, Burek turned into Alaska Air Cargo, backed up to a loading dock, and parked the van. "It's surprising how often they can't find the carcass," she said. We went inside. Burek handed a tracking number to an agent behind the counter. A man driving a forklift soon appeared at the loading dock. The forklift was freighted with a 31-gallon blue Rubbermaid Roughneck Tote labeled UNKNOWN SHIPPER. Burek opened the hatchback of the minivan and pushed aside pairs of Xtratuf rubber boots. The tote weighed a lot, but not so much that one man couldn't lift it.

We drove east through the sunny noonday traffic of Anchorage with a dead baby moose in the rear of the minivan. Burek was in a good mood, as she usually was. Years of working in close proximity to death had resulted in a sort of over-the-fence neighborliness with the macabre. She told me how area hospitals occasionally helped her determine cause of death by performing CT scans of dead baby orcas or by putting the heads of juvenile beaked whales into their MRI machines to look for acoustic injuries from Navy sonar or energy exploration. "I'm surprised this car doesn't smell worse for all the things that have been in it," she said. "I had a bison calf delivered to me, and it was in a tote like that, but it didn't fit—so these four legs were sticking out."

We arrived at a lab at the University of Alaska Anchorage, where Burek is an adjunct professor. The room was small, with white walls, a steel table at the center, and a drain in the floor. Burek pulled on a pair of rubber Grundens crabbing bibs the color of traffic cones, stepped into the tall boots from the minivan, and pulled her hair back. She could have been headed for a day of dipnetting for sockeye on the Kenai. An assistant laid out tools.

A big pair of garden shears sat on the counter, as foreboding as Chekhov's gun on the mantel.

"You're probably gonna want to put on gloves for this," she said.

Turned out of the tote onto the steel table, the moose calf was the size of a full-grown Labrador. It lay with its legs folded, as if it was just bedding down in soft lettuce. Burek flipped the calf onto its left side, which was how she liked to work on her ruminants. Then she began, calling out information. Sex: female. Weight: 74 pounds. Death: July 13. Length: 116 centimeters. Axillary girth: 76 centimeters. She swabbed an obvious abscess, open and draining, on the right shoulder. She noted the pale mucous membranes. She inserted a syringe into an eyeball to sample the aqueous humor. She returned to the shoulder, to the painful-looking abscess, and removed a piece of it for later examination on a glass slide under a microscope. Then Burek pressed her fingers into the wound.

"Oh, that's kind of gross," she said. "There's a comminuted fracture in there." When not using a scalpel and forceps, Burek often uses her fingers. After years of practice, her touch serves almost as a caliper and gauge. She will bread-loaf a liver and pinch the sections, probing for hardness. She will run her fingers along a wet trachea in search of abnormalities. "Oh, feel that," she will say to anyone willing to feel that.

Burek cut deeper to expose the wound. "Oh. *Oh.* Poor thing. It probably got nailed," she said. The detective was hitting her stride now. Searching the exterior of the calf, Burek quickly found what she was looking for—a second puncture wound, this one also badly infected. She measured the distance between the wounds: 5.5 centimeters, or the approximate distance, she estimated, between a bear's canines. "So that's cool."

In an interesting coincidence, Burek later would tell me she suspected that, for all the other abnormalities she found inside 13—the clot, the weird-looking lung—perhaps the otter, too, was ultimately done in by something as mundane as a predator. Blunt-toothed young killer whales will sometimes grab otters but not kill them, she explained; they sort of play with their food. Burek had seen it before. Intrigued, she telephoned the Museum of the North in Fairbanks and asked colleagues to measure the skull of a juvenile orca for an estimate of the diameter of its bite. The measurement perfectly fit the damage. "Of course, we'll never know for sure," she said. Still, there was a trace of satisfaction in her voice.

Now, using a No. 20 scalpel, Burek quickly skinned the moose calf and opened the stubborn clamshell of the rib cage. An unwelcome visitor wafted into the room. Burek, however, no longer seemed to notice odors that, were they canistered and lobbed across international borders, would swiftly be outlawed by the Geneva Conventions. As she worked, the gore took on a practiced orchestration. Burek cut triangles of beet-colored liver and dropped them into prelabeled bags with a pair of medical tweezers. She took samples of lung and lymph node and gall bladder. She squeezed the descending colon and collected the pellets. She filled vials and syringes. Some of the bits she did not even bother to label; after decades, Burek could recognize them by sight. With a few slices, she opened the firm dark knot of the heart like a chapbook and removed what resembled red chicken fat. At home Burek would spin the stuff in a centrifuge. Stripped of its red blood cells, the clear blood serum was an excellent way to see which infectious agents the animal had been exposed to in the past. "Diagnostic gold," Burek called it.

The table took on the appearance of a Francis Bacon canvas: A smear of blood. An ear divorced from a head. The sprung cage of the moose's body exposing its soft, translucent clockworks. The open mouth mutely horrified. Burek noted a hemorrhage on the surface of the pancreas and fibrin on the peritoneal cavity, and she moved on. The door of the lab stood open to the smiling July afternoon. Sunlight caught on aspen leaves. One of the two young women who were assisting Burek had just returned from her first year of veterinary school. Burek was her inspiration, she said. As the women laughed and worked, Burek quizzed her on biology and she told stories.

"I had a horse head in my freezer one time."

"Bears smell absolutely horrible. I did a bear necropsy in our garage once, and my son Thomas said I could never do that again."

"Can I get some muscle?"

"Those large whales? Holy cow. It's so confusing: where the heck is the urinary bladder?"

"For a while, I had a big colony in my garage of those flesh-eating beetles that museums use to clean skulls. But a couple of the beetles got out. That's when Henry put his foot down."

"Where'd my duodenum tag go? Anyone seen it?"

"I don't think rumens smell that bad. But I went to vet school in Wisconsin."

The steel table slowly emptied. The blue Rubbermaid bin filled. In went a foreleg. Intestines. The ear.

Now another assistant lifted the garden shears. She squeezed and sliced through the ball joint of the calf's femur, which is one of the best places on a young animal for Burek to see evidence of troubles, such as rickets, that would affect its growth plates. Burek, meanwhile, opened the skull to sample the brain.

"It's a bit of a mystery," she said as she worked, meaning the cause of the moose's death. Her initial guess: the bite led to septicemia, which led to encephalitis. "It's a story that kind of makes sense," she said. "I'd like to see more pus." Later she added: "But in this job you have to be willing to look dumb and be wrong and change your story."

Burek asked for the time. When I told her it was after four o'clock already, her good humor slipped. "I've got to get to the dump."

What was the hurry?

There was a new movie she wanted to see at seven, she said. She would have to race home to shower—to wash off the day, to wash off the smell, the blood, the moose.

"It's a Disney movie, I think," Burek said. A film about animals run amok. "It's called *The Secret Life of Pets*." She loaded the moose in the back of the minivan and reached for the bleach. "It looks cute."

PART V

"I'm Not a Mad Scientist—I'm Absolutely Furious"

Political Science

KAYLA WEBLEY ADLER

Female Scientists Report a Horrifying Culture of Sexual Assault

FROM *Marie Claire*

LET ME TELL you about my trouble with girls. Three things happen when they're in the lab: You fall in love with them, they fall in love with you, and when you criticize them, they cry." That's what British biochemist and Nobel laureate Tim Hunt told an audience at the World Conference of Science Journalists just two years ago.

Following intense social media backlash, Hunt claimed his remark "was intended as a lighthearted, ironic comment." His defenders said the line was taken out of context, pointing to his praise of female scientists in the same speech ("science needs more women, and you should do science, despite all the obstacles"). Bad joke or not, it cost Hunt his faculty position at University College London—and pointed to an insidious problem in the world of scientific research that had persisted far too long already.

From her office at the University of Illinois at Urbana-Champaign, Kate Clancy, Ph.D., watched the reaction to Hunt's comments unfold on Twitter, thinking, *Same shit, another day.* The thirty-eight-year-old associate professor was recalling a string of recent high-profile incidents that have made her and other women in science feel excluded, unwelcome, and like the very nature of being a woman in science "is in some way problematic," as she puts it. There was the peer reviewer for a scientific journal who suggested that two female researchers find "one or two male biologists" to coauthor and strengthen their paper. And the column published on *Science* magazine's website in which a biology professor told a postdoc who asked how to handle her adviser, who frequently tried to look down her shirt, to "put up with it, with good humor if you can."

In October 2015, Buzzfeed reported that University of California, Berkeley's Geoff Marcy, a noted astronomy professor, got away with sexually harassing students for at least a decade—despite complaints being filed against him at two different universities. (Marcy denied the allegations generally, but he apologized and has since resigned.) In January 2016, Democratic congresswoman Jackie Speier aired on the House floor a 2004 investigation into renowned astronomy professor Timothy Slater by his former employer, the University of Arizona, in which Slater was revealed to have gifted a student a vibrator, told a female employee she'd "teach better if she did not wear underwear," hosted meetings at strip clubs, and asked graduate students for sex. (Slater denies all allegations and is suing Arizona for releasing the confidential report.) A month later, *Science* magazine reported that paleoanthropologist Brian Richmond, curator of human origins at the American Museum of Natural History in New York City, was under investigation for sexually assaulting a research assistant. (Richmond said that the encounter in question was "consensual and reciprocal"; following multiple investigations by the museum, Richmond resigned last year.) Similar cases of harassment by science professors have also been reported at the California Institute of Technology, University of Chicago, University of Rochester, and Boston University.

Each time one of these incidents comes to light, and each time women in science take to social media to express empathy and anger, commenters swarm to tell them to get over it or learn how to take a joke—as if putting up with sexism, harassment, and assault is the expected price of being a female scientist. But Clancy knows it's never just a joke, or a few bad apples, or one misguided university—and she has the data to prove it.

In July 2014, Clancy and her coauthors Robin Nelson, Ph.D., Julienne Rutherford, Ph.D., and Katie Hinde, Ph.D., published the results of a survey that gauged the climate for women and men on field sites, where students and faculty at all levels across many scientific disciplines perform research. Of the 666 scientists from 32 different disciplines who completed the survey, 64 percent said they had personally experienced sexual harassment in the field and 20 percent reported having been sexually assaulted. Women were 3.5 times more likely to have experienced harassment or abuse than men, and were primarily harassed and assaulted by superiors (whereas male students were more likely to be victimized by peers). The report sent shock waves through the scientific community. "There was a lot of outrage," says Heather Metcalf, director of research and analysis at the Association for Women in Science. "A lot of, *How is this still happening? After all this time and all of these efforts to change the culture, how can this still be so prevalent?*" And worse: How many would-be female scientists are we losing as a result?

But Clancy and her coauthors knew that if they were truly going to impact the culture of their industry, quantitative research alone wouldn't be enough. The four are back this month with a follow-up study in the *American Anthropologist* journal detailing just how and why such widespread abuse has persisted. They conducted in-depth interviews with 26 of the original survey respondents and identified patterns and themes: on field sites with clear codes of conduct and supervisors who enforced the rules, women thrived, but on the sites where rules did not exist or were ambiguous and there were no consequences for wrongdoers, they found instances of unwanted flirtation or physical contact and intimidation, verbal sexual advances, sexist jokes and comments about physical appearance, forced kissing, attempted rape, and rape. One respondent said field site leaders insisted on conducting conversations while naked. Another said the head of her field site "would systematically prey on women" to the point that some women in her group chose to sleep on the floor in the same room rather than their own beds: "I had to serve as a kind of bodyguard."

Confronting offenders did not deter the perpetrators' behavior. Quite the opposite: women were only rewarded (i.e., given the best research assignments) if they consented to harassment or sexual advances. "It became clear that I was going to have to play along a little bit," another respondent said of her harasser. "I had a professional connection with this person, but he expected me to become his next mistress." Many female scientists said the behavior continued even after they left the field sites and that the psychological trauma compromised their ability to revisit, analyze, and publish their data. Several of the women were able to pin-point exactly how their abuse led to their career stalling. Five of the women said they had to leave the sciences altogether.

When she was a Ph.D. student studying paleoanthropology, Jenni-fer (not her real name) got the opportunity of a lifetime: she was invited to spend a month at a field site in Eurasia where she'd have the chance to study fossils crucial to the understanding of human evolution. One night, she and the other scientists shook off a long day of excavation at a house on-site—drinking, dancing, singing, listening to music. Jennifer was sitting next to the male principal investigator (or PI, as they're commonly called) who ran the site.

At some point, the lights went off. Almost as if on cue, the PI grabbed her and shoved his tongue in her mouth. "I don't want to say he French-kissed me because that's something two people do together," Jennifer says. "I felt filled up." She pushed him away but was sure everyone in the room had witnessed the encounter. Embarrassed, she feigned tiredness and scurried to bed.

For the next few weeks, the PI regularly grabbed her hand un-der the lunch table before she could swat him away. "I would just kind of politely laugh it off," she says. "I was out of my country, I didn't speak the language, and he was the head of my project. I hardly knew him at all—I only knew him as the person in charge of all my hopes and dreams."

Those dreams effectively ended the night of the incident. Be-fore that, Jennifer had planned to return to the site often, for as long as six months at a time, to conduct paleoanthropological re-search. But that PI was the gatekeeper of the site, and going back would have meant enduring further harassment and assault—he'd emailed her when she got home, saying how much he missed her and couldn't wait until she returned. "I knew it was over," she says. "I couldn't work with those fossils at all anymore—that's 6 million years of human evolution. Those fossils are why I went to grad school."

Jennifer was one of the lucky ones. She had to pivot her area of focus after four years of graduate school, which delayed her Ph.D., but she was able to continue her research using fossils in Kenya at a site free of harassment. Today, she's an associate professor of anthropology at a university in New England, but she's never stopped thinking of other women who may not have been as fortunate. "It gnaws away at me," she says. "I worry that it's happened to other women; I wonder how many of them just gave up."

It's stories like Jennifer's that made Clancy dig deep into the darkest corners of her field. Back in 2011, she was blogging for *Scientific American* and was often invited to speak at universities about "how to be a woman in science." At one such event, she ran into an old colleague. The friend used to be about two years behind Clancy in her career, but when they reconnected, the woman had yet to finish her Ph.D., while Clancy was well into a tenure-track job. Clancy asked what was holding her up. "Every time I look at my dissertation data it reminds me of when I was sexually assaulted in the field," the woman replied. A few weeks later, Clancy had a remarkably similar conversation with another woman at a different conference. "I was like, 'What's going on with you?'" Clancy recounts. "And she was like, 'Let me tell you about the systemic sexual harassment at the field site that made me completely desert this line of research.'"

The women agreed to write about their experiences anonymously on Clancy's blog. When the posts were published, the comments and emails from women who had also experienced gender discrimination and harassment on field sites—everything from always being the person designated to do the dishes to being raped by their PIs—came rolling in. Clancy wondered just how many female scientists' careers had been stymied by sexism and assault.

She asked Nelson, Rutherford, and Hinde if they would be willing to study the disturbing trend with her. The women agreed but worried that doing such work as junior faculty was risky for their own careers. They considered waiting to conduct the survey until they had all secured tenure but decided they couldn't "in good conscience" wait. "If there is a penalty for trying to make things better and safer, then why would I want to be in this field?" explains Hinde, an associate professor of human evolution at Arizona State University, of her rationale. "Of course I was scared, but this is too important."

Clancy presented the initial results at the April 2013 meeting of the American Association of Physical Anthropologists. "I was terrified to stand up in front of a bunch of my colleagues and basically say, 'A hefty portion of you are harassing and assaulting your students and your peers. We're watching you now and we're not going to let you pull that shit anymore.'"

The reason those men were able to get away with it for so long? So few women were sounding alarms. According to the research, called Survey of Academic Field Experiences (SAFE)—eventually published in *PLOS ONE* after being rejected twice, which the women suspect was on account of the subject matter—only 13 percent of female survey respondents said they reported the harassment, and just 7 percent of women who were sexually assaulted formally came forward.

That's likely because budding female scientists have an extreme disincentive to report—doing so often means speaking out against the very people who write their letters of recommendation and fund their work. The perpetrators "are the people who have control over our careers," says Nelson, an assistant professor of anthropology at Santa Clara University. "There's no way for me to overstate how devastating reporting harassment or assault could be. These people are so powerful, and so the idea of bringing charges . . . I mean, why? You would sabotage your whole career."

A majority of the dozen victims interviewed by MarieClaire .com for this story—most of whom did not want their names published for fear of retribution—said that when they reported, the perpetrators were rarely punished. They often felt as though the person who victimized them was untouchable. Not only did their perpetrators have tenure (meaning they can only be fired for "extraordinary circumstances"), but their work brought the university millions of grant dollars and immeasurable prestige. "We have to stop privileging the science above the people who are doing it," Hinde says.

One woman, a biology undergraduate student at a small Midwestern college, told MarieClaire.com that one of her professors grabbed her breast three times while they were reviewing an exam in his office. She reported the incident to the administration, but the professor wasn't punished—only *her* reputation suffered. "I'm no longer the smart girl who likes plants and trees," she says. "I'm the girl who got groped and made trouble for her teacher." When one twenty-seven-year-old planetary scientist reported being harassed and stalked by a visiting scholar at the observatory where she worked, the head of the department told her to "grin and bear it," she recalls. "He said, 'You wouldn't want to complain about him because it could ruin his career.'" No such concern was expressed for her own.

The toll sexism takes on a woman's career in the sciences isn't always so overt. Studies have shown there are many subtle ways women are hindered: female scientists receive less mentoring, are given fewer awards, are invited to speak at fewer conferences, and receive significantly less funding for their research. They are also less likely to be hired—in 2012, a Yale University researcher sent otherwise identical applications to faculty for lab manager jobs and found "John" was more likely to be given the position and receive a higher starting salary than "Jennifer." Research has also shown that male professors see female students as less competent and are significantly less likely than female faculty to bring women trainees into their labs. "You can tell women, 'Lean in, fight for yourself, fight for a raise, work hard, put in extra hours,'" Clancy says, "but at the end of the day, these are the kinds of things that are actually going to hold us back."

In 2009, a woman we'll call Sarah was working as a research assistant in a lab where she was studying for a Ph.D. in neuroscience. One night, she was driving back to campus from a work event with the married head of her lab when, out of nowhere, he started talking about having sex with her. "It was really graphic and gross—he suggested not using a condom so I could feel the ridge of the head of his penis," she recalls. "I was like, 'Um, I'm just going to drop you off at your car.'" She hoped that would be the end of it.

Sarah decided she would go to work the next day and act as though nothing had happened. When she was getting ready, her boss called. He asked her not to mention to anyone what had occurred. Then he said, "Don't worry about coming in—you can take the day off." So she stayed home. From then on, any time she did go to work, her boss was "just so awkward" and kept telling her not to come in. "I felt unwelcome, which is weird because he should be the unwelcome one," Sarah says. "But it was his lab."

She went from working 20 hours each week to putting in 5 hours or less. Soon, "it was like, 'Well, we started this project, but you are not a part of it. We are writing up this paper, but we didn't put you on it because you haven't been here,'" Sarah remembers. By the end of her time there she'd only been listed as an author on 4 papers, while her male counterpart had been cited on 20.

Over time, it became clear that she wasn't going to get the job she'd wanted since first grade. "It was a sad, sad moment," she says. "I had been investing so much of myself for so long, and then to have to say, 'Well, that was all down the drain . . .'" Today, Sarah works at a pharmaceutical company as a liaison to doctors. "I don't do research—I just talk about other people's research," she says. "I do enjoy it, but it's not what I dreamed of. I'm no longer a scientist."

We may never know just how many potential female scientists like Sarah we're losing. But we do know this: between 2014 and 2015, women obtained 59 percent of bachelor's degrees in the biological sciences and 38.5 percent of bachelor's degrees in the physical sciences. (Tell that to anyone who claims women just aren't interested in science.) But while women graduate in equal numbers, they don't make it to the highest ranks in equal numbers. Only 21 percent of professors across the sciences are female. According to the Census, male science and engineering graduates are employed at twice the rate women are.

But for all the downsides, the good news is that, thanks to Clancy and her colleagues, these issues are finally being discussed and reform is on the way. It's too soon to say what measures may come from their follow-up study, but already a number of disciplines—including archeology, astronomy, physics, and ecology—have conducted or are planning similar surveys specific to their research sites. Most are finding comparably high levels of misconduct. In response, organizations that fund research, like NASA and the National Science Foundation, have made public statements that they will not tolerate harassment at grantee institutions and programs. The American Association of Physical Anthropologists drafted a new code of ethics. A group of professors at universities nationwide is working to develop sexual harassment bystander intervention training for the earth, space, and environmental sciences.

Change may also come at the legislative level. Representative Speier has introduced a bill requiring schools to share information about disciplinary proceedings so that when perpetrators move colleges, their indiscretions follow them. "It's a thrill to feel like we are really making a difference," says Rutherford, an associate professor in the department of women, children, and family health science at the University of Illinois at Chicago. "I hear from people who say, 'I can now have this conversation with the people who are in a position to change things, and they have to respect it because it's on paper in black and white.'" Hinde adds that she's heard from colleagues who have implemented sexual harassment policies at their field sites and has had "men email me and tell me they had marginalized someone in the past and they aren't going to do it anymore."

That's key, Rutherford says, because if the culture doesn't change, we'll never know what might happen if female scientists are allowed to reach their full potential. "What if women didn't have to be so vigilant all the time? What if you didn't have to smile, have a thick skin, keep it to yourself, look over your shoulder, or be careful about who's in the lab at the end of the night," she posits. "How much more could women achieve?"

RACHEL LEVEN

A Behind-the-Scenes Look at Scott Pruitt's Dysfunctional EPA

FROM *Center for Public Integrity*

ENVIRONMENTAL PROTECTION AGENCY administrator Scott Pruitt doesn't hide his contempt for how the agency has been run but does profess to care about one of its key programs: Superfund, which oversees the cleanup of the nation's worst toxic-waste sites. In April, he toured a site in East Chicago, Indiana, contaminated with lead and arsenic, and told residents, "We are going to get this right."

The following month, Pruitt—Oklahoma's attorney general before he joined the EPA—tapped one of his former donors, banker Albert "Kell" Kelly, to find ways to accelerate and improve Superfund cleanups. Kelly started by consulting career staff members—often-knowledgeable officials who work at the agency regardless of who holds the White House. But then Kelly closed off the process, conferring with Pruitt to produce a final plan that altered or excluded many of the staffers' suggestions. Gone, for example, was the idea that EPA officials be identified early on to lead discussions with communities on how contaminated land should be used after cleanup.

"We're missing a huge opportunity to do something new and different with Superfund," said one of two EPA employees who described the process to The Center for Public Integrity on the condition of anonymity.

What happened with Superfund is hardly an anomaly. Today's EPA is wracked with internal conflict and industry influence, and is struggling to fulfill its mission, according to more than two dozen current and former agency employees. A few dozen political appointees brought in under the Trump administration are driving policy. At least 16 of the 45 appointees worked for industries such as oil, coal, and chemicals. Four of these people—and another 21 —worked for, or donated to, politicians who have questioned established climate science, such as Pruitt and Senator James Inhofe, R-Oklahoma.

Career staff members—lawyers, scientists, analysts—are largely being frozen out of decision-making, current and former agency employees say. These staffers rarely get face time with Pruitt and frequently receive top-down orders from political appointees with little room for debate. They must sometimes force their way into conversations about subjects in which they have expertise.

And that is a big mistake, said one of Pruitt's predecessors.

Career employees are "very dedicated to protecting human health and the environment, and they will change their ways of how they do that if they're convinced you really want to accomplish that aim," said Christine Todd Whitman, EPA administrator under President George W. Bush.

One such employee agreed. "I think it's the fact that we're not following regular procedures, we're not sure of what the legal justification is for some of the things they're asking us to do. We're just kind of being told 'Do the opposite thing you did 18 months ago.' That's hard to swallow."

The EPA staffers who spoke to the Center say the isolation of Pruitt's top staff from the rest of the agency limits the perspectives the administrator is exposed to before making decisions. Two appointment calendars, covering a six-month period beginning in March, show that Pruitt hears overwhelmingly from industry. He was scheduled to meet 154 times during the period with officials from companies such as ExxonMobil and trade associations such as the American Petroleum Institute, the oil industry's biggest lobby group. API was among at least 17 donors to Pruitt when he ran for state or federal office or led the Republican Attorneys General Association that have met with him as EPA administrator. Those same calendars indicate he saw only three groups representing environmental or public-health interests, though an EPA press release says he met with two others.

EPA spokeswoman Liz Bowman disputed claims that the roughly

15,000-person agency is riven with discord. Career employees are "vital" to the EPA's work and meet regularly with political appointees and the administrator, she said, citing ongoing deliberations on a water-pollution rule as an example. (Two career employees who spoke to the Center at Bowman's request confirmed that they routinely work with political appointees and did so on the water rule.)

"We talk to people throughout the regions, the states, the career staff, and a variety of different perspectives prior to making decisions," Bowman said, adding in an email that "we follow the Administrative Procedure Act in our regulatory process, meaning taking into consideration comments submitted by ALL commenters, including environmental NGOs, the public, and other commenters."

For the public, much is at stake. Under Pruitt, who sued the EPA 14 times as Oklahoma's attorney general, the agency already has declined to ban a pesticide linked to neurological damage in children, frozen requirements to reduce water pollution from coal-fired power plants, and opened the door to loosening limits on toxic coal waste. The EPA most recently proposed eliminating the Clean Power Plan, an Obama administration rule aimed at reducing carbon emissions in the power sector.

"These rules [being rolled back] aren't perfect by any stretch of the imagination. There are ways to improve things," said Gordon Binder, who served as chief of staff for then–EPA administrator William Reilly under President George H. W. Bush. "But Pruitt's come in with a flyswatter and is slapping them down instead of laying out the problems with a rule and saying, 'How can we fix it?'"

Who's Running the EPA?

There's a striking absence of high-level leadership at the EPA. Only 1 of 13 positions requiring Senate confirmation—Pruitt's—is filled. These positions—higher-ranking than those held by the 45 political appointees—include leaders of key agency offices that oversee air, water, and other programs. Six people have been nominated for the 12 open slots and are awaiting confirmation (although two of them have joined the EPA in the meantime as senior advisers).

Most major decisions are made on the third floor of the William Jefferson Clinton South Building on Pennsylvania Avenue, where Pruitt and his handful of confidants have their offices. Among those who most prominently have Pruitt's ear: EPA chief of staff Ryan Jackson, who previously held the same position with Inhofe, and Samantha Dravis, who worked with Pruitt at the Republican Attorneys General Association and now leads the EPA's Office of Policy.

Dravis, widely seen as Pruitt's closest adviser, came to the agency with a background in law and politics but little environmental experience. That's a departure from her two immediate predecessors, who were already at the EPA when they were tapped to lead the policy office.

Jackson helped Inhofe negotiate major bipartisan deals in the Senate, such as the passage of federal chemical-policy reform. Still, emails released under the Freedom of Information Act to *The New York Times* show that Jackson directed career staff to deny a petition seeking to ban chlorpyrifos, a pesticide suspected of harming children's brains, directly contradicting the recommendations of EPA scientists.

Bowman said in an email that career staff were "instrumental" in drafting the denial. The former acting head of the EPA's chemicals office, a career employee who left the agency last month, told the *Times* she opposed the decision, even as she followed Jackson's instructions.

Other influential political appointees include Sarah Greenwalt, senior adviser to the Office of the Administrator, who worked for Pruitt when he was Oklahoma's attorney general; Bowman, head of public affairs, who came from the American Chemistry Council, a chemical-industry trade group; and Mandy Gunasekara, a senior policy adviser who worked on the Republican staff of the Senate Environment and Public Works Committee.

Nine career staff members told the center their opinions seem to hold little weight. They are excluded from meetings, they say, and their advice on agency operations is often disregarded. Some believe this is because of the flurry of leaks that have come from inside the agency since Pruitt took office. Political appointees have lashed out at suspected leakers and relieved them of work assignments, even in the absence of proof, career employees said.

"They are terrified of career staff leaking," one said. "And once they get an idea in their head [about] someone, they won't change it."

Bowman said, "Those concerns have not been brought to our attention. And if they are we will do everything we can to address them."

Pruitt has broken with tradition by foregoing many introductory briefings with career staff designed to help new administrators set priorities, several current and former employees said. Instead, he's worked to roll back EPA rules, an effort that also diverges from common practice.

Political appointees are taking more of a hands-on role in tasks career employees previously would have handled. Take, for example, a recent notice announcing EPA plans to reconsider whether certain vehicle-emission standards for greenhouse gases were too strict. Career staff had drafted a concise version of the notice, but appointees expanded the number of vehicles affected by the review and made the Department of Transportation the lead agency on the decision, despite the EPA's legal obligations to control planet-warming emissions under the Clean Air Act.

"This was a much more major rewrite" than would have happened under previous administrations, said an EPA employee familiar with the matter. "At least one plausible outcome of this process," the employee said, "is that the EPA would unilaterally abdicate its [legal] responsibility."

"Checking the Box"

On the rare occasions when career employees are asked to brief Pruitt, he seems unwilling to change his antiregulatory posture on major industry priorities, according to some career employees. Betsy Southerland, who headed an office within the EPA's water program until August, said her team met with Pruitt twice about a rule designed to limit wastewater discharges from power plants as he considered weakening parts of it.

The team told Pruitt that industry arguments against the rule already had been considered and found to be inaccurate. They offered more nuanced actions the administrator could take to address concerns expressed by the Small Business Administration, a separate federal agency, and the Utility Water Act Group, a lobbying organization. Pruitt was unmoved, she said. In August, the EPA announced it would reconsider key parts of the rule.

"You get the feeling that his mind was made up before we started the briefing process," Southerland said. "It looked like he was kind of checking the box to meet with us." The *Times* found similar behavior by Pruitt when the EPA declined to ban chlorpyrifos, Pruitt didn't follow agency scientists' advice, having "promised farming industry executives who wanted to keep using the pesticide that it is 'a new day and a new future,'" the *Times* reported.

These episodes could simply reflect inexperience—appointees struggling to figure out the agency they lead. They also could reflect pressure placed on the EPA by executive orders and presidential memoranda to act quickly on big-ticket issues like the Clean Power Plan.

But Pruitt's pro-industry bent has convinced some current and former employees that he and his like-minded advisers are aiming to destroy the EPA from the inside.

"Look, I think he does not support what the agency has been trying to do for 40 years," said William Ruckelshaus, EPA administrator under Presidents Richard Nixon and Ronald Reagan. "He wants to dismantle—not improve or reform—the regulatory system for protecting public health and the environment."

PART VI

"It's Not Rocket Science . . . (Actually, Some of It Is)"

Space Science

REBECCA BOYLE

Two Stars Slammed into Each Other and Solved Half of Astronomy's Problems. What Comes Next?

FROM FiveThirtyEight

PROGRESS, AS THEY say, is slow. In science, this is often true even for major breakthroughs; rarely is an entire field of research remade in a single swoop. The Human Genome Project took a decade. Finding the first gravitational waves took multiple decades. So it's hard to overstate the enormous leap forward that astronomy took on August 17, 2017.

On that day, astronomers bore witness to the titanic collision of two neutron stars, the densest things in the universe besides black holes. In the collision's wake, astronomers answered multiple major questions that have dominated their field for a generation. They solved the origin of gamma-ray bursts, mysterious jets of hardcore radiation that could potentially roast Earth. They glimpsed the forging of heavy metals, like gold and platinum. They measured the rate at which the expansion of the universe is accelerating. They caught light at the same time as gravitational waves, confirmation that waves move at the speed of light. And there was more, and there is much more yet to come from this discovery. It all happened so quickly and revealed so much that astronomers are already facing a different type of question: now what?

"Even people like me, who have been waiting for this for a long time and preparing for this, I don't think we're ready," said Edo Berger, an astronomer at Harvard who studies explosive cosmic events. "Now it's a question of, do we have the right instrumentation for doing all the follow-up work? Do we have the right telescopes? What's going to happen when we have not just one event, but one a month, or one a week—how do we deal with that flood?"

From Copernicus and Kepler to Hubble and Einstein, astronomy has experienced plenty of tectonic shifts. The discovery of GW170817, as the August event is known, will be another of these. Astronomers often describe the detection of gravitational waves —which happened for the first time just last year, and was awarded a Nobel Prize in October—as a new form of perception, as though we can now hear as well as see. The neutron-star merger event was like seeing and hearing at the same time, and with a dictionary to make sense of it all.

The August 17 gravitational wave gave astronomers a glimpse at an entirely different universe. For most of history, they've studied stars and galaxies, which seem static and unchanging from the vantage point of human timescales. "You can look at them today and look at them 10 years from now, and they will be the same," Berger said. But GW170817 revealed a universe *alive,* pulsating with creation and destruction on human timescales. Think about that: GW170817 was a relatively close 130 million light-years from Earth, meaning its gravitational waves and light were emitted while the first flowering plants were busy evolving on Earth, around the time stegosauruses roamed the plains. But the event itself unfolded in less than three human-designated *weeks.* This faster timescale is "pushing the way astronomy is done," Berger said.

When the wave crashed through Earth, it caused a tiny shift in the path of laser beams traveling down long corridors in observatories called the Laser Interferometer Gravitational-Wave Observatory (LIGO), in the United States, and the Virgo interferometer, in Italy.* On August 17, LIGO's twin detectors and Virgo each felt the wave, which allowed astronomers to roughly triangulate from which direction it rolled in. They swung every bit of glass they had,

---

* These machines are so sensitive that even the burbling of coffee brewing in another building can confuse the detectors, but the signals they're searching for are so weak that even a strong gravitational wave signal, like one produced by two colliding giant black holes, only nudges the beam by a fraction of the width of a *proton.*

both on Earth and in the heavens, in that general direction. In space, the Fermi space telescope glimpsed a burst of gamma radiation. Within an hour, astronomers made six independent discoveries of a bright, fast-fading flash: a new phenomenon called a kilonova. Astronomers saw the telltale sign of gold being forged, a major discovery by itself. Nine days later, X-rays streamed in, and after 16 days, radio waves arrived too. Each type of information tells astronomers something different. Richard O'Shaughnessy, an astronomer at the Rochester Institute of Technology, describes the discovery as a "Rosetta stone for astronomy." "What this has done is provide one event that unites all these different threads of astronomy at once," he said. "Like, all our dreams have come true, and they came true now."

As O'Shaughnessy put it, every discovery eventually becomes a tool. Astronomers hope to use neutron-star mergers to test general relativity, the mind-bending conceit that what we perceive as gravity is actually a curving of space and time. Binary neutron stars and black holes may deviate from the gravitational fields predicted by general relativity, which could put Einstein—and alternative theories for gravity in extreme systems—to the test, said Jacqueline Hewitt, a physicist who directs MIT's Kavli Institute for Astrophysics and Space Research.

Gravitational waves aren't blocked by dark matter, dust, or other space objects, so they can serve as messengers from the insides of stars, Hewitt said. When LIGO upgrades are finished next year, astronomers will be able to investigate how the waves form and reconstruct the violent smashups.

Eleonora Troja, an astronomer at NASA's Goddard Space Flight Center who studies X-rays, had hoped for years to detect the light from a neutron-star merger, but many people thought she was dreaming. "I had a lot of proposals rejected because they were considered too visionary, too advanced," she said. But even Troja never imagined witnessing what happened this summer. "Sometimes, I am still like, 'Did that really happen?'"

Troja says that the information gathered in August could eventually serve as a template for finding other neutron-star collisions and gamma-ray jets. We may have already unwittingly captured evidence of many such events, but the record is likely buried in a decade's worth of data from the Fermi and Swift gamma-ray space telescopes, waiting to be uncovered. Those observatories, and new

ones under construction now, will allow humanity to see even
more violent, rapidly changing astronomical phenomena. The
Large Synoptic Survey Telescope, for example, is currently under
construction and will eventually photograph the whole sky every
three nights. "In the future, when we digest all this information, it
will be a drastic change in the way we study these cosmic objects,"
Troja said.

This event that unfolded across a couple of weeks will also inform our deepest experience of time, the beginning and the end
of our cosmology. Combinations of light and gravitational waves,
like those detected after the neutron-star merger, can be used to
measure the rate at which the expansion of the universe is accelerating.*

"It's totally new," Troja said. "Comparing the two independent
measurements, the one from light and the one from gravitational
waves, you can measure the rate of the expansion of the universe."
All our futures are wrapped up in this question.

Thanks to the August 17 event, astronomers now know what to
look for. Soon, they will be able to sift through an embarrassment
of neutron-star mergers and other phenomena. And as with any
disruption, there will be a period of adjustment. Huge telescopes
in space and on Earth are few and far between, and on Earth, most
of them can only work when it's dark and the skies are clear. That
means thousands of people vie for limited time at the proverbial
eyepiece. Telescope committees are set up to review proposals and
grant that time, and assignments are often allotted months in advance. That will have to change as astronomers chase events in
real time.

"In our case, for the telescopes we were using in Chile, people
traveled to Chile to use the telescopes, and we asked them to give
up their time [to track the August 17 event]. And everybody did it
with so much enthusiasm," Berger said, adding that anyone who
sacrificed hard-won telescope time was credited in the scientific
literature. "But we need better mechanisms. You can't call up ev-

---

* Astronomers currently do this using "standard candles," which include variable
stars that flash like lighthouses, as well as supernovas, which are the death throes
of huge, exploding stars. The idea is that if you know the standard brightness of an
object, and you can tell how bright it seems to you versus how bright it's supposed
to look, that can tell you how far away it is and how fast it's going. In a similar way,
neutron-star mergers can serve as "standard sirens."

ery individual person and negotiate and explain, especially when these objects are fading away so quickly, while you're on the phone with them."

Hewitt is chair of a committee that develops 10-year plans for astronomy, known as the decadal surveys, and said the detection of gravitational waves—and neutron-star mergers—were listed as goals for the next several years in the most recent report in 2010 and in the midpoint report in 2015. We got there early, and now astronomers are talking about how to prioritize their time, where to focus, and how to pivot to the next big thing, she said. Many are now hoping that the United States rejoins a space-based gravitational wave experiment called LISA. And they are talking about how to turn their eyes to the sky, at a moment's notice, the next time the universe throws something big their way.

"It's a wonderful time, it's a terrifying time," O'Shaughnessy said. "I can't really capture the wonder and the horror and glee and happiness."

KENNETH BROWER

The Starship or the Canoe

FROM *California*

IN 2015, AN observatory high in the Atacama Desert of Chile
detected three planets orbiting an M star, an ultra-cool dwarf,
in the constellation Aquarius about 40 light-years, or 232 trillion
miles, from Earth. Until then, the dim star was designated 2MASS
J23062928–0502285. Not such a charming name. The discoverers
of its satellites, a team of astronomers who operate the Chilean
observatory remotely from Liège in Belgium, took the opportunity
to warm up that appellation. The Belgians called the new system
TRAPPIST-1, after the robotic telescope that had done the work,
TRAPPIST, the Transiting Planets and Planetesimals Small Telescope.

This acronym cheats a little with the gratuitous "I," a nod to
the Trappist order of monks, who have several famous abbeys in
Belgium. Inmates of the abbeys produce wonderful beers, dark,
full of residual sugars and living yeast, beers that improve with
age, like wine. The astronomers celebrated their find with monk-
brewed bottles of Westvleteren 12, the best beer in the world.

Last September, NASA's Spitzer Space Telescope, with help
from several land-based telescopes, verified two of the planets
spotted by the Belgians and went on to detect five more, for a total of seven, the biggest batch of Earth-sized planets ever detected
around a single star.

When my editor at *California* proposed that, in light of this jack-
pot, I revisit an old book of mine, *The Starship and the Canoe*, I
agreed on the spot. Would I like to apply the dialectic of my book,
written 40 years ago, to the seven new exoplanets? Yes, I would.

Three of seven are in the "Goldilocks zone," not too hot for life, not too cold, just right. An ideal temperature for asking again a question posed by my book: To what should we be adapting? To this blue-green sphere down here, with its single sun, good for only 5 billion more years, or to the glittery firmament above?

The Starship and the Canoe is an account of two vessels, two contrary views of our future, and two men, a father and son, hell-bent on voyaging in opposite directions. The father was, and is, Freeman Dyson, the stellar astrophysicist, particle physicist, and mathematician, a prodigy drawn since infancy to the stars, author of his first papers on planetary mechanics at the age of five, before he really got the hang of spelling. Most of his subsequent career he has spent at the Institute for Advanced Study in Princeton, with sojourns in places like Berkeley and the Max Planck Institute in Germany.

Dyson's most unusual sabbatical came in 1958, when he moved to La Jolla to become a luminary—first magnitude—on a team of 50 scientists and engineers working on Project Orion, a scheme to explore the solar system by "nuclear-pulse propulsion." Atomic bombs would drop from the Orion spacecraft like eggs from a duck. The bombs would detonate beneath a massive "pusher plate" at the bottom of the vessel, and enormous shock absorbers would smooth out the jolts. The team studied Coca-Cola vending machines for ideas on how to dispense their nuclear explosives. They were not keen on handing their spaceship over to any crew of astronauts. They wanted to go themselves. Their plan was to bomb themselves to Mars by 1965 and Saturn by 1970.

It was totally insane, of course, except apparently not. The Air Force and ARPA, the Advanced Research Projects Agency, funded Orion, and for a time the project held its own against the chemical rocketry of Wernher von Braun. No conceptual flaw killed Orion. What vaporized the project, finally, was the Partial Test Ban Treaty of 1963, which forbade nuclear explosions in the atmosphere.

The goal of the Orion effort was the design of a modest little ship for touring this solar system, but in one lecture, early on, Freeman Dyson strayed way outside the box. The physicist stunned his colleagues by pushing the Orion idea as far as it could go. What would it take, he asked himself, to send Chicago to Alpha Centauri, the nearest star system? What it would take, he calculated,

was 30 billion pounds of deuterium loaded into 25 million ther-
monuclear bombs and racked in a spidery ark, a colossal "heat-
sink" starship, with a base 12 miles in diameter, that would accelerate for a century and then glide on to Alpha Centauri, reaching
that system, in the Orion Arm of the Milky Way, in a little more
than 150 years. The cost, as Dyson figured it, was reasonable. We
could perpetuate humanity in the stars for about one Gross National Product.

George Dyson, Freeman's only son, as a preschooler was enthu-
siastic about trips like this, but lost interest in his teens.

George was drawn less to galaxies than to the woods. He was
an explorer of New Jersey swamps, a collector of snakes and turtles. Helen Dukas, Einstein's personal secretary, babysat him. His
father's mentor, the Nobel laureate Hans Bethe, would drop in.
Professor Bethe won his Nobel for figuring out the chemistry that
fuels the brightness of the stars, and then headed up the Theoretical Division of the Manhattan Project, where he helped guide that
experiment to its dazzling conclusion, and then went on to stoke
the small, brief sun of the H-bomb. "Bright eyes," he remembered
of George as a small boy underfoot. That was the only feature he
could recall for me. Enormous theoretical firepower gravitates
to the Institute for Advanced Study, but it was mostly wasted on
George. He was a smart boy, but what gave him most satisfaction
was working with his hands. His pastime from an early age was
building model boats.

George's parents divorced when he was six. After the breakup,
he spent most of the year with his father in Princeton and often
summered with his mother. The Dysons, father and son, had trouble understanding each other. George grew his hair out longer
than his father liked. When he was twelve he built his first kayak in
his room. Denied use of the garage, he was forced to extend construction into his closet, like an old salt building a clipper ship in
a bottle. It was hell getting the kayak out again—a sore point with
George for years afterward.

When George was fourteen, a package arrived for him from
California and was intercepted at the post office. He had spent the
summer with his mother, the mathematician Verena Huber-Dyson,
who was teaching at Berkeley, and he had discovered the Haight-
Ashbury. The package contained marijuana. George was arrested
in class and led away in handcuffs. "That really raised my status

around there," he says, but admits that it also dropped his spirits. His father thought he should stay in jail a while, so as to learn his lesson, and this led to what would become a six-year estrangement.

At sixteen, George enrolled at UC San Diego. He didn't like it, so after a few weeks he transferred to the Berkeley campus, where he didn't like it either. Skipping class, he found himself strolling the Berkeley Marina past a sailboat with a "For Sale" sign. He bought it, spending almost all his allotment for school expenses. He moved into his boat, cooked in the galley on a Primus stove, and slept on one of his four narrow berths. When he studied, which was infrequently, he liked to sail to Angel Island and anchor in Richardson Bay while he read.

Two summers earlier in the Sierra Nevada, he had worked as a pot-washer for my sister Barbara, who was cook on a Sierra Club trip. ("Twenty-five dollars a week," he recalled recently. "My first big break in life.") On arriving at Cal, he looked Barbara up, and now and then, when he needed company or his ship's larder ran low, he visited our Berkeley house. He and my little brother, John, who are the same age, became very close, de facto brothers. My brother Bob, an anti-industrialist, was president of an imaginary company he had founded, the Strange Development Corporation, all its products conceptual. Seeing promise in George, he appointed him vice president. My father, then the executive director of the Sierra Club—an actual organization—overheard George's mission statement and would quote it in speeches for the rest of his life: "To find freedom, without taking it from someone else." My mother, then editor for the Department of Anthropology, came across George once in the kitchen reading the ingredients on a bag of dog food. He had been policing the shelves for additives. "This is the only thing in the house that's fit to eat," he muttered.

George lasted just one term at Berkeley. Selling his boat, he moved up to British Columbia and took up residence in a Douglas fir outside Vancouver, building a tree house 95 feet above the planetary surface. The house incorporated 14 branches as structural members and was lashed to the tree, not nailed, for the treetop swayed in the wind and the attachment had to be flexible. The single small room, shingled inside and out with red-cedar shakes he split from a drift log, was narrower than George is tall. His bed, Procrustean, had octagonal windows at head and foot. There was a small wood-burning stove. Working on boats in the labyrinthine

straits and sounds of the Northwest coast, George went barefoot
often and tanned so dark that he was sometimes taken for an Indian. He dedicated himself to resurrecting one of the finest of
indigenous North American skin boats, the Aleut kayak—the *baidarka,* the Russian sea-otter hunters called it. His first big effort was
a 30-foot *baidarka* with cockpits for three. As fieldwork for my book
on the Dysons, I paddled that vessel with George from Glacier Bay
in Alaska down to Canada. Then in 1975, he took the idea "kayak"
as far as it could go. He built an analogue to his father's starship,
a 48-foot canoe with six holes for paddlers, *Mount Fairweather,* the
biggest kayak in the world.

TRAPPIST-1 is an auspicious acronym, given my assignment here.
Planetesimals, represented by the second P, happen to be Freeman
Dyson's favorite heavenly bodies, the most promising extraterrestrial spots for human habitation, in his view. And the whole of the
acronym has a nice echo of George. Trappist monks live in isolation, eschew idle talk, eat a vegetarian diet, and make their own
bread and beer. These are the Rule of Saint Benedict. A nearly
identical code, homegrown, guided the laconic young man with
whom I tried to start up conversation on our kayak trips. In the
tight little monastic cell of his tree house, rocked to sleep by the
flex of his treetop in the wind, George was Trappist in all but
the vows.

There was something for me, too, out there in Aquarius, a kind
of delayed synchronicity. The telltale light of 2MASS J23062928-
0502285, the dwarf star—the very photons which, by blinking as
the seven planets transited the dwarf's face, revealed the existence
of the group—was emitted 40 years ago, even as I, half a lifetime
and 232,000,000,000,000 miles away, tapped out, on a quaint machine called a typewriter, the final sentences of *The Starship and the
Canoe.*

The NASA webpages on the new solar system open with an
artist's conception of the surface of the sixth planet, TRAPPIST-
1f. The painting is beautifully done. It belongs in the genre of
cover art for science-fiction novels, except that there is no spacecraft, or alien, or scantily dressed space maiden in the picture. A
channel of open water—or open liquid of some other molecule
—leads out between icebergs across a wine-dark sea to the horizon, where the dwarf sun sets. The sky is spitting snow or sleet. (If

future colonists are lucky, this will be H_2O snow, not frozen flakes of CO_2, such as fall on the southern hemisphere of Mars.) The scene looks very Antarctic, except that the ultra-cool dwarf, in its nearness, looms four times larger than our sun and burns much paler and cooler.

Next comes a 360-degree view from an imaginary spot on the surface of TRAPPIST-1d. You scroll horizontally past black, striated boulders—basalt, from the look of them—embedded in flats and berms of pale sand. Having come full circle on the planetary surface, you can scroll vertically down into the sand at your feet or up into the hazy red sky.

This is all NASA moonshine. If we know anything with absolute certainty, it is that none of these TRAPPIST worlds will look anything like the depictions. Every planet and moon in our own system, once we have come to know it intimately, has proven to be a complete surprise, entirely its own world, unlike any other. Why should it be any different out in Aquarius?

"The discovery is a huge piece of luck," Freeman told me of TRAPPIST-1.

The physicist, now ninety-three, continues to follow closely all the news from the stars, as he has done for the past ninety years.

"Luck that there happens to be a very small star so close to us with the planets lined up precisely in the plane where we can see them transiting. But it is not only luck. The people who planned the observations knew what they were looking for. The smaller the star, the bigger the fraction of light that each planet will obstruct, and the more precisely the planets are seen and measured. I was excited by the discovery as a triumph of intelligent observation."

The seven planets make a cozy system, orbiting at such close quarters that the surface details of some would be visible to the naked eye from others. (Cozy but dizzying, as a year in this incestuous whirl of sister planets lasts between just 1 and 20 Earth days.) One might guess that Freeman would be charmed by this system as a place to colonize. I doubted it.

"First I have to clear away a few popular misconceptions about space as a habitat," he told a London audience in a 1972 lecture. "It is generally considered that the planets are important. Except for Earth, they are not. Mars is waterless, and the others are for various reasons inhospitable to man."

Much more promising, Dyson said, were the comets, small worlds a few miles in diameter, rich in water and other chemicals essential to life. Thousands of millions of comets, loosely attached to the sun, await us out there, by his estimate, with a combined surface area a thousand or ten thousand times greater than Earth's. Closer in, in the big gap between Mars and Jupiter, is the belt of the asteroids, the rubble of collisions between two or three planets no longer there, except as fragments. There is water in the asteroids. They should make prime real estate, in Freeman's opinion. Ceres is 593 miles in diameter. Pallas is 319 miles wide. Eros, in outline roughly the same shape and twice the size of Manhattan, tumbles through space like a dead cigar or a severed whatever. Icarus, a boulder a mile in diameter, flies awfully close to the sun, but otherwise might make a congenial outpost for some physicist's reclusive son or some other type of hermit.

"I am not impressed by the hype about the Goldilocks zone," Freeman said, when I asked. "The purpose of astronomy is to look at everything in the universe and find all kinds of unexpected mysteries. Of course an Earthlike planet is especially interesting because the Earth has such a rich history and geography. But there is no strong reason to expect life to be confined to any Goldilocks zone. Life as we know it is wonderfully adaptable, and Nature's imagination is richer than ours."

What does impress Freeman is the volume of information that space telescopes like the Spitzer and the Kepler are sending back to us. Both these infrared telescopes are now busy monitoring thousands of stars in the constellations Cygnus and Lyra. It is the Spitzer telescope that verified the last four TRAPPIST planets.

"This find is only one item in a huge treasure of discoveries made by missions like Kepler," Freeman said. "Kepler observed with high precision the variations of light from individual stars. The public only hears about variation caused by transiting planets. There is at least as much variation caused by internal processes in the stars. For most professional astronomers, stars are more interesting than planets. There will be many more missions like Kepler in the future. To me, the great news is that we are at the beginning of a new era in the exploring of the universe, with a far more detailed understanding of both stars and planets. The seven-planet system is like the panda in the San Diego Zoo, rightly admired as beautiful and as a public attraction. But the zoo as a whole is more important than the panda."

In summer of 1975, in my VW camper, I drove Freeman and his daughter Emily from the Nanaimo Ferry 150 miles northward through the forest of Vancouver Island to Kelsey Bay, where we took another ferry to Beaver Cove for a reunion with George. Freeman did not mind my coming along. He and his son had been five years apart, and he thought that having a third party on hand, as intermediary, made sense.

"The big moment," he said, as the ferry reversed engines and the water churned against the pilings of the slip. I spotted George walking down toward the waterfront: knit cap, stiff oilskin jacket, familiar gait. I pointed him out to his family, who had yet to know him as a grown man. George, for his part, didn't recognize us until I had driven off the ferry and pulled up alongside. Beaming, he and his father shook hands vigorously and kept at it for a long time.

Over the next days, we visited George's acquaintances in the maze of glaciated islands at the foot of Queen Charlotte Sound. First we crossed to Swanson Island, which for the moment had just one inhabitant, George's friend Will Malloff. The other half of the population, Will's wife, Georganna, a sculptress, was temporarily off island. Malloff is inventor, among other things, of the Alaska mill, which is in widespread use across North America by people returning to the land. He had come to Swanson Island four years before with $50 and a chainsaw. Since then he had built, or salvaged from elsewhere, a whole settlement: house, greenhouse, sheds, chicken coop, duck pen, pheasant aviary, and separate workshops for himself and Georganna, whose sculptures stood everywhere in the clearing and deeper in the forest.

Malloff and the other inhabitants of this temperate rainforest fascinated Freeman. They demonstrated, he believed, the hardiness and resourcefulness that will be required of pioneers in the comets or stars. To the amusement of his son and everyone else, he tried to recruit George's people as space colonists. His special target was Malloff. One day we watched from shore as Malloff waded out in his gum boots, bent over his Mercury outboard, and commenced repairing its blown head gasket. Reaching for the Vise-Grip pliers in his back pocket, he had a thought.

"I won't let him send me into space unless I can take my Vise-Grips!" he swore, gesturing with the pliers.

"I can't send you," said Freeman. "You have to want to go yourself."

Watching George and Freeman together on the island, I was struck anew by their close resemblance. George is a taller version of Freeman. Both men are lean, with large noses previously broken. Both laugh a characteristic Dyson laugh in which the shoulders shake but no sound comes out. Both have piercing eyes they hold wide open. (George's mother noticed, early on, this thing with the eyes. In her diary entry for February 10, 1954, when her son was not quite one year old, Verena Huber-Dyson wrote, "George sometimes seems to have Freeman's unicorn look in his eyes. That glimpse of Thurber's unicorn reaching out for the lily simply haunts me.")

Where I had thought my book on the Dysons would be a simple story of opposites, it was proving much more complicated and interesting, a story also of convergences. Both Dysons believed in small, creative societies (which Freeman had found in his Orion team, and George in places like Swanson Island). Both wanted to get far away. Both craved a fresh start.

If in George's teen years, the trajectories of the two Dysons were headed straightaway toward diametrically opposed vanishing points, then by Swanson Island their arcs were beginning to come back around. The curvature of space-time, or whatever is bending this process, would appear to be sharper than Einstein calculated; it has now brought the Dyson worldviews closer than I would ever have imagined back then. George today still has a kayak workshop, but it is mostly idle. He is now a historian of science and the author of books on the evolution of technology: *Darwin Among the Machines,* and *Turing's Cathedral,* and *Project Orion.* The barefoot days are over. At his TED Talks he wears shoes.

In March of 2009, George and his older sister, Esther Dyson, an entrepreneur, investor, and philanthropist, had traveled with Freeman to a former Soviet missile base in Kazakhstan for the launch of a Russian rocket, a Soyuz mission to the International Space Station.

"Esther invested in a company in Russia called Yandex," George explained. "She's actually one of the founders. It's the only thing

Esther ever did that made a big pile of money. Most of her investments, she's made a million here or there, but in Russia she helped found this search engine, it became the Google of Russia, and then it went public and she got a pile of rubles. She couldn't take it out of Russia, or she didn't want to take it out. Esther, she's never driven a car, she doesn't own a house. So she thought, 'What can I do with a whole pile of Russian money?' She decided to buy a seat on one of those Russian rockets.

"It's an irony. When we were kids, Freeman was building a space ship, and I *really* wanted to go. I was five years old, and I thought they were building it there in San Diego. I imagined that the big, round building at General Atomic was the launch pad for the spaceship, and we were going to get in it and go. Then all these years later, it's actually Esther. The story of my life! Everything I ever did was always outshadowed by Esther. So we end up going to Kazakhstan. Esther was the backup on the rocket flight. Every time they fly one of these tourists, they have a backup, in case you break your leg the week before the flight. She took all the training. Survival course, and so on. Even the standby seat costs $3 million."

The missile base, it seemed to George, was caught in a time warp, stuck in the Sputnik era. "They haven't changed anything," he said. "It worked in 1960, and that's the way they still do it. The telephones are all dial phones. The whole launch site is run on coal-fired generators. The place looks like it was built by Soviet slave labor. Which it was."

Charles Simonyi, the billionaire for whom Esther was understudy, and who had paid $35 million for his seat, did not break a leg, unfortunately. "He brought his beautiful young Swedish wife and all her family," said George. "They were wearing mink coats out in the Kazakh desert. And Esther shows up with her motley family. They put us all up in the Sputnik Hotel. Everything was very choreographed. It was like a wedding. Russian Orthodox priests. One side of the family were aristocrats, and the other side were hillbillies."

The rocket went up, with the three Dysons, the hillbillies, watching from the ground. George and Freeman had a whispered conversation. They agreed that this was crazy.

"The real irony is that here I am with Freeman, 50 years after Project Orion, watching the American pay $35 million to go into *very* low Earth orbit on a Russian rocket. It's just so depressing. None of those space dreams came true. Or, they totally didn't come true in terms of manned space flight. They absolutely did come true in terms of the robot machines, which are sending back all these incredible pictures."

In 1975, Freeman brought reading material to Swanson Island. Among the books in his bag was an annotated version of *Slaughterhouse-Five,* Kurt Vonnegut's fictionalized, black-humor account of the firebombing of Dresden, which the novelist witnessed from the ground as a German POW. This was the first Vonnegut book I had seen given this sort of academic gloss.

I had no doubt back then, nor do I now, why Freeman was giving the firebombing of Dresden this close read between the lines. In July 1943, at the age of nineteen, he had reported for duty as mathematician at the headquarters of the Royal Air Force Bomber Command. Just days thereafter, on July 27, British bombers succeeded, for the first time in history, in igniting a firestorm, burning 42,000 Germans to death in Hamburg. On every big raid after that first one, the Allies tried for another firestorm, but the bombers succeeded only twice more in Europe—Hamburg again and then Dresden. "We still do not understand why," he recalled. "It probably happened when there was some instability in the atmosphere before the attack so that meteorological energy amplified the energy of the fires. The firestorms had a big effect on civilian casualties." Targeting civilians is a war crime. Freeman felt complicit.

Interviewing him in 1975, with the embers of Dresden 30 years cold, I asked whether firestorms figured in his drive for the stars. He nodded. "Our strongest feelings are subconscious. I grew up in a time of despair—the late '30s. It was far worse than it is now." In a postwar essay, Freeman wrote, "In my personal view of the human situation, the exploration of space appears as the most hopeful feature of a dark landscape."

Slaughterhouse-Five is Vonnegut's bleakest book, too haunted for me. My own favorite is his second, *The Sirens of Titan.* At the heart of that one is a parable that better illuminates the question at hand: colonization, or not, of outer space.

All events in this novel unfold from a strange accident suffered by one of its characters, Winston Niles Rumfoord, a New England aristocrat wealthy enough to build his own spacecraft. (An Elon

Musk or Jeff Bezos sort of figure, or the other way around, as Vonnegut conceived him years before the birth of either of those entrepreneurs.) Two days out of Mars, Rumfoord steers into an uncharted space hazard, a chrono-synclastic infundibulum. *(Chrono* is Greek for time. *Synclastic* means curving all to the same side, like the skin of an orange. *Infundibulum* is Latin for funnel.) Swept up by this funnel, like Dorothy by the tornado, Rumfoord and his great hound, Kazak, become wave phenomena, twin spirals stretching from origins in the sun to termini in the star Betelgeuse. Their spirals intersect at regular intervals with Earth or Mars or Mercury, at which times Rumfoord and Kazak materialize briefly on those planets. Titan, Saturn's largest moon, is the only spot in the universe where they remain solid and corporeal year round.

Chrono-synclastic infundibula, it turns out, are commonplace in the universe. They are everywhere, wild threads in the worn fabric of space-time. The infundibula are Vonnegut's metaphor for the huge surges of radiation, the bone loss in astronauts, and all the unimagined hazards waiting out there in the dark matter and the light. The moral of the infundibula, their message to humanity, he writes, is simple:

What makes you think you're going anywhere?

What does make us think we're going anywhere? Colonization of the stars is wonderful as romance, but is up against two intractable realities: the cosmological and the biological. The cosmic problem is distance. The TRAPPIST-1 system, at 40 light-years and 232 trillion miles away, is indeed *relatively* close, yet still unimaginably far. It is nearly 40 times more distant than Alpha Centauri, which Freeman's heat-sink starship would have taken 150 years to reach.

The biological problem is the ephemerality of species. *Homo sapiens* is just 200,000 years old, yet already we are on our way to becoming something else. How much time do we have left, *as us,* to complete some of these interminable crossings? What sort of creature will shamble out when the hatch opens on the other side? (In *The Sirens of Titan,* Winston Niles Rumfoord addresses this difficulty: "What an optimistic animal man is!" he cries. "Imagine expecting the species to last for ten million more years—as though people were as well-designed as turtles!") Evidence mounts that we have passed our peak and a devolutionary downslide has begun. Consider the GOP. Just 150 years ago it was the party of Lincoln.

The word *adaptation* came into wide use after Darwin, and its most common application is still biological, a reference to the adjustments that an organism makes to conditions in its ecosystem if it is to survive. A species is not just its DNA. A species is an interaction between its genes, its culture, and the other organisms and climate of a particular place. Every instant of our evolutionary adaptation, to date, has been in the community of life on this third planet from the middling G star we orbit. So far as we know, to date, there is no community of life out there with which we might continue our coevolution. Our manifest destiny, it seems obvious to me, is down here on Earth.

In visiting the TRAPPIST planets illustrated on the NASA website, I grew increasingly uneasy with the artists' conceptions. I began noticing doubtful details. For example: In the waves at the foot of the icebergs on TRAPPIST-1f new ice is forming. The ice panes are polygons of sharp angles floating slightly separated, like the pieces of a finished jigsaw puzzle after the dog has bumped the table. It could be that the seas on TRAPPIST-1f are methane, like the lakes on Titan. It could be that methane ice really does behave this way near an M star. But it is also possible that the artist was unfamiliar with how ice forms in our own cold seas—first as slush and then pancake ice, worn circular, with raised rims all around, from collisions with other pancakes in the waves.

For any conceptual artist rendering a new dinosaur, there are fossilized bones to work from, and sometimes the tracks of that reptile, and educated guesses to be made from reconstruction of its ecosystem. For the illustrators of TRAPPIST-1, there is only periodic dimming of light from an ultra-cool dwarf. In that signal there is enough information to suggest planetary density and allow speculation that the seven planets are rocky. Beyond that, the illustrations are figments.

These made-up planets are fun. They spur imagination in viewers. They make, or pretend to make, elusive astronomical realities palpable. As PR for NASA, they jack up public interest, keep appropriations flowing in, and fund new missions. But they are false. They suggest that we know more about these worlds than we do. How many viewers, I wonder, even those who have noted the "artist's conception" advisory, come away believing that the look of these planets is based on actual evidence.

My killjoy instincts in this are annoying even to me. Their source, I think, is in the first manned mission to circumnavigate another world.

On Christmas Eve of 1968, Bill Anders, an astronaut orbiting the lunar surface in Apollo 8, snapped a picture through the module window: Earthrise above the dead white curvature of the moon. It is commonplace for environmentalists to mark this as the most important photograph ever taken, and legions of commentators from other disciplines have marked *Earthrise* as the epiphany of epiphanies. The Apollo 8 astronauts certainly saw it that way. "We came all this way to explore the moon," said Bill Anders, "and the most important thing is that we discovered the Earth."

There it was, our place, a small blue marble, a "cloudy," kids call it An oasis in the blackness of space.

A NASA illustrator has borrowed this motif for the agency's website, painting the TRAPPIST-1 planets as varicolored marbles in a line against the blackness of the void. TRAPPIST-1b looks like Mars with acne. TRAPPIST-1c looks like just regular Mars. TRAPPIST-1d has a longitudinal chain of turquoise lakes suspiciously similar to Grand Prismatic Springs in Yellowstone. TRAPPIST-1e and TRAPPIST-1f are both Earthlike, bluish and wreathed in cloud. TRAPPIST-1h looks like Venus.

These marbles are all mirage. The Apollo 8 photograph *Earthrise* is a true picture of a real world and our actual circumstances.

Thousands of exoplanets have been discovered in the past two decades, and artistic liberties have been taken with many. To the extent that this artwork plants the idea, even subconsciously, that we have alternatives, that these far places are real possibilities, that we can solve our pollution and population problems somewhere out in Aquarius or the Orion Arm, or even closer in, by mining the asteroids or colonizing comets, then it subverts the epiphany of *Earthrise*. The solutions to our problems will be down here.

In the early 1980s, NASA announced a plan to send a journalist into space on the shuttle. *Reader's Digest* believed it had the inside track in placing a writer on board. The magazine liked *The Starship and the Canoe*, believing that the book's duality—the view both Earthward and spaceward—was what it wanted, and it proposed that I go. Unmarried then, with no kids, I agreed on the spot. In writing up my résumé, I exaggerated my physical fitness, which was already good, and I glossed over problems I have with machinery and instrument panels.

It did not escape me that if, indeed, I made the cut for the shuttle, then Freeman was the man who had bought me my ticket. It was ironic. In constructing *The Starship and the Canoe*, I had been careful not to let my own biases infect the whole book. I had confined my opinion to a chapter titled "My Opinion." In that one I threw in with George and his canoe. I came out against Freeman and his starship. ("Freeman has faith in the hospitality of space, and I have faith, almost equally groundless, in its inhospitality.") What a hypocrite. The first time someone offers to send me into orbit, I jump.

There was a setback. NASA decided, for incomprehensible reasons, to send a teacher first. Christa McAuliffe, who taught high school in New Hampshire, won out over 11,000 candidates. By the day of the *Challenger* mission that launched her, January 28, 1986, I had a five-week-old son. The baby had diminished somewhat my enthusiasm for risking my life on assignments, yet I was envious still of McAuliffe. She had jumped her place in line.

Like tens of millions of Americans, I watched the televised liftoff and then, 73 seconds later, the explosion, pieces of the ship rocketing off in all directions, with one giant smoky contrail curving back upon itself. Like everyone, I had a moment of confusion and denial. Was this fireball just the first booster disintegrating? Then came the keening moan from the crowd at Cape Canaveral and the shock and grief. Within that emotion, for me, was a small, shameful countercurrent of relief, as in a passenger whose missed train has cost him his berth on the *Titanic*.

What made me think I was going anywhere?

SUSANNAH FELTS

Astonish Me: Anticipating an Eclipse in the Age of Information

FROM *Catapult*

THERE ARE THEORIES all over YouTube. Already I know too much. My special glasses meet the Transmission Requirements of ISO 12312–2, Filters for Direct Observation of the Sun. My nine-year-old returns home from school with key vocabulary words set to the tune of that old 1980s ballad "Total Eclipse of the Heart." If my head wasn't already stuffed with facts, she could provide.

But really, I'd like to be less ready. On the afternoon of August 21, what I want is to be astonished.

In the thick haze and cicada drone of this afternoon, the moon's shadow will race at 1,800 miles an hour across the United States, its path like a pageant sash draped from Oregon shoulder to South Carolina hip. On that decorated continental body the city of Nashville, my home, will be the largest to experience totality. This area hasn't seen a total eclipse for more than 500 years; not since 1979 has the United States experienced one. Annie Dillard wrote ecstatically of that planetary pas de deux, and now her words surface repeatedly in my Facebook feed, shared for the occasion. But while Dillard's 1979 eclipse darkened the Pacific Northwest and stretched into Canada, this one will shadow only the Lower 48—a first-ever occurrence—and sweep neatly from coast to coast (the next eclipse to do that will happen in 2045). For an estimated 55 million Americans, a swath of Tennessee will provide the closest and best view. You can imagine the nervous murmurs about traffic.

In early July, the viewing parties began to twinkle, distant stars on the blank expanse of the calendar, all clustered upon this one August Monday. A city-sponsored gathering at the baseball stadium; viewing parties at local parks. A brunch at a fancy hotel, with eggs Benedict and waffles. Free hot dogs in an auto dealer's parking lot. Farm parties with celebrity chefs and live music.

I want to make the most of it, but I do not want to sip mimosas among strangers. I do not want to pay for parking or get a free commemorative koozie. I don't want to feel, first, pity, for the eclipse-swag vendors—what a way to spend the celestial event of a lifetime, hawking souvenirs—and then shame, because who am I to say that padding one's wallet with the profit from trinkets manufactured overseas isn't just the right way to do the Great American Eclipse of 2017?

Still, I can't stop thinking about the fact that, centuries ago, a total eclipse was cause for panic, a sign of doom—or a moment of great discovery.

Today, it is the strangest of things: absolutely predictable and astoundingly rare. The thought of it ignites my curiosities, my fierce devotion to the magic of the natural world. Just the phrase *path of totality* charms me—the repetition of sounds, the metaphoric sinew. Its very definition—the area covered as the moon's shadow, or umbra, hits the Earth—demands I slow down and consider it. A journey made by darkness, a trail 70 miles wide that will vanish as quickly as it appeared. On a hill in the Yakima Valley, as the sky went dark, Annie Dillard heard the other eclipse watchers scream. I might wish to feel that kind of awe, or fear.

An eclipse is also the surest of things, no alarms and no surprises. While so much of life—the very moments that will alter us the most—we cannot predict, we can know even the minute truths of a spectacular celestial event with utter certainty.

Yet it is a happening so far from the infinitesimal movements of us humans that to break it down to our science almost seems a hubristic offense. How can I restore to it some measure of its power to surprise? How can I make my two minutes count?

The answer, in the end, is simple. I know a place, a half hour from downtown and ideal for eclipse-viewing. Large, wide-open lawn, a dome of sky above. Home to sundry animals and a small planet's worth of rare plant species. Bourbon on the shelf, baby goats frisking in the barn, hummingbirds buzzing about—until the dark suddenly quiets them all. This, my parents' land, is where

I have decided to watch the sky go dark and feel the air go cool. My family beside me, all of us growing a little older as the shadow hits us and we listen to the birds cease to sing.

It will be my parents' last eclipse; it will be my daughter's first. It may well be the only total eclipse any of us are granted. In their presence, I hope to kindle the awe that too much knowledge, or the flurry of public celebration, might just stamp out. I try to imagine what I will remember, on normal days far into the future, of that 1 minute and 57 seconds. "Usually it is a bit of a trick to keep your knowledge from blinding you," Dillard writes. "But during an eclipse it is easy. What you see is much more convincing than any wild-eyed theory you may know."

Maybe she's right, and it will be easier than I can even imagine. At the very least, the rest of today will fall away, and what remains will be the people, the place, the quiet luck of proximity and the even quieter mysteries of our interior selves, both so chromosomally shared and so separate. Two minutes, a lifetime, the overlapping of lives. Our threaded-together fates still largely unknown.

I want to hold my child, our heads tilted to the heavens, the very opposite image of gazing down at our devices. We will marvel at Dillard's "old wedding band in the sky, or a morsel of bone."

And when the blackness peels away and August torpor clamps down again (August, I do not care for you, it's true) — what will we carry forth? I hear my daughter, wanting to know what we're going to do next. My husband: "Back to work." Me, checking this off my to-do list: *eclipse*. None of us will be spooked right to death. "One turns at last even from glory itself with a sigh of relief," Dillard writes. "From the depths of mystery and even from the heights of splendor, we bounce back and hurry for the latitudes of home."

Yet I write this wishing, impossibly, for what only fiction could bestow upon those scant two minutes: for life to irrevocably change — at the very moment when we seem vividly lost in our darkest legacies. In the absence of such wizardry, I will accept the stillness and awe experienced alongside the ones I love most. I will accept a summer day turned inside out.

And I write this wanting to know: How did you and you and you experience the eclipse? Whose hands did you clasp at the wonder of the moon blotting out the sun, breaking the day in two?

STEVEN JOHNSON

Greetings, E.T. (Please Don't Murder Us.)

FROM *The New York Times Magazine*

ON NOVEMBER 16, 1974, a few hundred astronomers, government officials, and other dignitaries gathered in the tropical forests of Puerto Rico's northwest interior, a four-hour drive from San Juan. The occasion was a rechristening of the Arecibo Observatory, at the time the largest radio telescope in the world. The mammoth structure—an immense concrete-and-aluminum saucer as wide as the Eiffel Tower is tall, planted implausibly inside a limestone sinkhole in the middle of a mountainous jungle—had been upgraded to ensure its ability to survive the volatile hurricane season and to increase its precision tenfold.

To celebrate the reopening, the astronomers who maintained the observatory decided to take the most sensitive device yet constructed for listening to the cosmos and transform it, briefly, into a machine for talking back. After a series of speeches, the assembled crowd sat in silence at the edge of the telescope while the public-address system blasted nearly three minutes of two-tone noise through the muggy afternoon heat. To the listeners, the pattern was indecipherable, but somehow the experience of hearing those two notes oscillating in the air moved many in the crowd to tears.

That 168 seconds of noise, now known as the Arecibo message, was the brainchild of the astronomer Frank Drake, then the director of the organization that oversaw the Arecibo facility. The broadcast marked the first time a human being had intentionally transmitted a message targeting another solar system. The engineers had translated the missive into sound, so that the assembled group would have something to experience during the transmission. But its true medium was the silent, invisible pulse of radio waves, traveling at the speed of light.

It seemed to most of the onlookers to be a hopeful act, if a largely symbolic one. a message in a bottle tossed into the sea of deep space. But within days, the Astronomer Royal of England, Martin Ryle, released a thunderous condemnation of Drake's stunt. By alerting the cosmos of our existence, Ryle wrote, we were risking catastrophe. Arguing that "any creatures out there [might be] malevolent or hungry," Ryle demanded that the International Astronomical Union denounce Drake's message and explicitly forbid any further communications It was irresponsible, Ryle fumed, to tinker with interstellar outreach when such gestures, however noble their intentions, might lead to the destruction of all life on Earth.

Today, more than four decades later, we still do not know if Ryle's fears were warranted, because the Arecibo message is still cons away from its intended recipient, a cluster of roughly 300,000 stars known as M13. If you find yourself in the northern hemisphere this summer on a clear night, locate the Hercules constellation in the sky, 21 stars that form the image of a man, arms outstretched, perhaps kneeling. Imagine hurtling 250 trillion miles toward those stars. Though you would have traveled far outside our solar system, you would only be a tiny fraction of the way to M13. But if you were somehow able to turn on a ham radio receiver and tune it to 2,380 MHz, you might catch the message in flight: a long series of rhythmic pulses, 1,679 of them to be exact, with a clear, repetitive structure that would make them immediately detectable as a product of intelligent life.

In its intended goal of communicating with life-forms outside our planet, the Arecibo message has surprisingly sparse company. Perhaps the most famous is housed aboard the Voyager 1 spacecraft—a gold-plated audiovisual disc, containing multilingual greetings and other evidence of human civilization—which slipped free of our solar system just a few years ago, traveling at a relatively sluggish 35,000 miles per hour. By contrast, at the end of the three-minute transmission of the Arecibo message, its initial pulses had already reached the orbit of Mars. The entire message took less than a day to leave the solar system.

True, some signals emanating from human activity have traveled much farther than even Arecibo, thanks to the incidental leakage of radio and television broadcasts. This was a key plot point in Carl Sagan's novel *Contact*, which imagined an alien civilization detecting the existence of humans through early television broadcasts from the Berlin Olympic Games, including clips of Hitler speaking at the opening ceremony. Those grainy signals of Jesse Owens, and later of Howdy Doody and the McCarthy hearings, have ventured farther into space than the Arecibo pulses. But in the 40 years since Drake transmitted the message, just over a dozen *intentional* messages have been sent to the stars, most of them stunts of one fashion or another, including one broadcast of the Beatles' "Across the Universe" to commemorate the 40th anniversary of that song's recording. (We can only hope the aliens, if they exist, receive that message before they find the Hitler footage.)

In the age of radio telescopes, scientists have spent far more energy trying to look for signs that other life might exist than they have signaling the existence of our own. Drake himself is now more famous for inaugurating the modern search for extraterrestrial intelligence (SETI) nearly 60 years ago, when he used a telescope in West Virginia to scan two stars for structured radio waves. Today the nonprofit SETI Institute oversees a network of telescopes and computers listening for signs of intelligence in deep space. A new SETI-like project called Breakthrough Listen, funded by a $100 million grant from the Russian billionaire Yuri Milner, promises to radically increase our ability to detect signs of intelligent life. As a species, we are gathered around more interstellar mailboxes than ever before, waiting eagerly for a letter to arrive. But we have, until recently, shown little interest in sending our own.

Now this taciturn phase may be coming to an end, if a growing multidisciplinary group of scientists and amateur space enthusiasts have their way. A newly formed group known as METI (Messaging Extra Terrestrial Intelligence), led by the former SETI scientist Douglas Vakoch, is planning an ongoing series of messages to begin in 2018. And Milner's Breakthrough Listen endeavor has also promised to support a "Breakthrough Message" companion project, including an open competition to design the messages that we will transmit to the stars. But as messaging schemes proliferate, they have been met with resistance. The intellectual descendants of Martin Ryle include luminaries like Elon Musk and Stephen Hawking, and they caution that an assumption of interstellar friendship is the wrong way to approach the question of extraterrestrial life. They argue that an advanced alien civilization might well respond to our interstellar greetings with the same graciousness that Cortés showed the Aztecs, making silence the more prudent option.

If you believe that these broadcasts have a plausible chance of making contact with an alien intelligence, the choice to send them must rank as one of the most important decisions we will ever make as a species. Are we going to be galactic introverts, huddled behind the door and merely listening for signs of life outside? Or are we going to be extroverts, conversation-starters? And if it's the latter, what should we say?

Amid the decommissioned splendor of Fort Mason, on the northern edge of San Francisco, sits a bar and event space called the Interval. It's run by the Long Now Foundation, an organization founded by Stewart Brand and Brian Eno, among others, to cultivate truly long-term thinking. The group is perhaps most famous for its plan to build a clock that will successfully keep time for 10,000 years. Long Now says the San Francisco space is designed to push the mind away from our attention-sapping present, and this is apparent from the 10,000-year clock prototypes to the menu of "extinct" cocktails.

The Interval seemed like a fitting backdrop for my first meeting with Doug Vakoch, in part because Long Now has been advising METI on its message plans and in part because the whole concept of sending interstellar messages is the epitome of long-term decision-making. The choice to send a message into space is one that may well not generate a meaningful outcome for a thousand years, or a hundred thousand. It is hard to imagine any decision confronting humanity that has a longer time horizon.

As Vakoch and I settled into a booth, I asked him how he found his way to his current vocation. "I liked science when I was a kid, but I couldn't make up my mind which science," he told me. Eventually, he found out about a burgeoning new field of study known as exobiology, or sometimes astrobiology, that examined the possible forms life could take on other planets. The field was speculative by nature: after all, its researchers had no actual specimens to study. To imagine other forms of life in the universe, exobiologists had to be versed in the astrophysics of stars and planets; the chemical reactions that could capture and store energy in these speculative organisms; the climate science that explains the weather systems on potentially life-compatible planets; the biological forms that might evolve in those different environments. With exobiology, Vakoch realized, he didn't have to settle on one discipline: "When you think about life outside the Earth, you get to dabble in all of them."

As early as high school, Vakoch began thinking about how you might communicate with an organism that had evolved on another planet, the animating question of a relatively obscure subfield of exobiology known as exosemiotics. By the time Vakoch reached high school in the 1970s, radio astronomy had advanced far enough to turn exosemiotics from a glorified thought experiment into something slightly more practical. Vakoch did a science-fair project on interstellar languages, and he continued to follow the field during his college years, even as he was studying comparative religion at Carleton College in Minnesota. "The issue that really hit me early on, and that has stayed with me, is just the challenge of creating a message that would be understandable," Vakoch says. Hedging his bets, he pursued a graduate degree in clinical psychology, thinking it might help him better understand the mind of some unknown organism across the universe. If the exosemiotics passion turned out to be a dead end professionally, he figured that he could always retreat back to a more traditional career path as a psychologist.

During Vakoch's graduate years, SETI was transforming itself from a NASA program sustained by government funding to an independent nonprofit organization, supported in part by the new fortunes of the tech sector. Vakoch moved to California and joined SETI in 1999. In the years that followed, Vakoch and other scientists involved in the program grew increasingly vocal in their argument for sending messages as well as listening for them. The "passive" approach was essential, they argued, but an "active" SETI —one targeting nearby star systems with high-powered radio signals—would increase the odds of contact. Concerned that embracing an active approach would imperil its funding, the SETI board resisted Vakoch's efforts. Eventually Vakoch decided to form his own international organization, METI, with a multidisciplinary team that includes the former NASA chief historian Steven J. Dick, the French science historian Florence Raulin Cerceau, the Indian ecologist Abhik Gupta, and the Canadian anthropologist Jerome H. Barkow.

The newfound interest in messaging has been piqued in large part by an explosion of newly discovered planets. We now know that the universe is teeming with planets occupying what exobiologists call "the Goldilocks zone": not too hot and not too cold, with "just right" surface temperatures capable of supporting liquid water. At the start of Drake's career in the 1950s, not a single planet outside our solar system had been observed. Today we can target a long list of potential Goldilocks zone planets, not just distant clusters of stars. "Now we know that virtually all stars have planets," Vakoch says, adding that, of these stars, "maybe one out of five have potentially habitable planets. So there's a lot of real estate that could be inhabited."

When Frank Drake and Carl Sagan first began thinking about message construction in the 1960s, their approach was genuinely equivalent to the proverbial message in a bottle. Now, we may not know the exact addresses of planets where life is likely, but we have identified many promising ZIP codes. The recent discovery of the TRAPPIST-1 planets, three of which are potentially habitable, triggered such excitement in part because those planets were, relatively speaking, so close to home: just 40 light-years from Earth. If the Arecibo message does somehow find its way to an advanced civilization in M13, word would not come back for at least 50,000 years. But a targeted message sent to TRAPPIST-1 could generate a reply before the end of the century.

Frank Drake is now eighty-seven and lives with his wife in a house nestled in an old-growth redwood forest, at the end of a narrow, winding road in the hills near Santa Cruz. His circular driveway wraps around the trunk of a redwood bigger than a pool table. As I left my car, I found myself thinking again of the long now: a man who sends messages with a potential life span of 50,000 years, living among trees that first took root a millennium ago.

Drake has been retired for more than a decade, but when I asked him about the Arecibo message, his face lit up at the memory. "We had just finished a very big construction project at Arecibo, and I was director then, and so they said, 'Can you please arrange a big ceremony?'" he recalled. "We had to have some kind of eye-catching event for this ceremony. What could we do that would be spectacular? We could send a message!"

But how can you send a message to a life-form that may or may not exist and that you know nothing at all about, other than the fact that it evolved somewhere in the Milky Way? You need to start by explaining how the message is supposed to be read, which is known in exosemiotics as the "primer." You don't need a primer on Earth: you point to a cow, and you say, "Cow." The plaques that NASA sent into space with Pioneer and Voyager had the advantage of being physical objects that could convey visual information, which at least enables you to connect words with images of the objects they refer to. In other words, you draw a cow and then put the word *cow* next to the drawing and slowly, with enough pointing, a language comes into view. But physical objects can't be moved fast enough to get to a potential recipient in useful timescales. You need electromagnetic waves if you want to reach across the Milky Way.

But how do you point to something with a radio wave? Even if you figured out a way to somehow point to a cow with electromagnetic signals, the aliens aren't going to have cows in their world, which means the reference will most likely be lost on them. Instead, you need to think hard about the things that our hypothetical friends in the TRAPPIST-1 system will have in common with us. If their civilization is advanced enough to recognize structured data in radio waves, they must share many of our scientific and technological concepts. If they are hearing our message, that means they are capable of parsing structured disturbances in the electromagnetic spectrum, which means they understand the electromagnetic spectrum in some meaningful way.

The trick, then, is just getting the conversation started. Drake figured that he could count on intelligent aliens possessing the concept of simple numbers: 1, 3, 10, etc. And if they have numbers, then they will also very likely have the rest of what we know as basic math: addition, subtraction, multiplication, division. Furthermore, Drake reasoned, if they have multiplication and division, then they are likely to understand the concept of prime numbers—the group of numbers that are divisible only by themselves and one. (In *Contact,* the intercepted alien message begins with a long string of primes: 1, 2, 3, 5, 7, 11, 13, 17, 19, 23, and so on.) Many objects in space, like pulsars, send out radio signals with a certain periodicity: flashes of electromagnetic activity that switch on and off at regular rates. Primes, however, are a telltale sign of intelligent life. "Nature never uses prime numbers," Drake says. "But mathematicians do."

Drake's Arecibo message drew upon a close relative of the prime numbers to construct its message. He chose to send exactly 1,679 pulses, because 1,679 is a semiprime number: a number that can be formed only by multiplying two prime numbers together, in this case 73 and 23. Drake used that mathematical quirk to turn his pulses of electromagnetic energy into a visual system. To simplify his approach, imagine I send you a message consisting of 10 X's and 5 O's: XOXOXXXXOXXOXOX. You notice that the number 15 is a semiprime number, and so you organize the symbols in a 3-by-5 grid and leave the O's as blank spaces. The result is this:

x		x		x
x	x	x		x
x		x		x

If you were an English speaker, you might just recognize a greeting in that message, the word *HI* mapped out using only a binary language of on-and-off states.

Drake took the same approach, only using a much larger semiprime, which gave him a 23-by-73 grid to send a more complicated message. Because his imagined correspondents in M13 were not likely to understand any human language, he filled the grid with a mix of mathematical and visual referents. The top of the grid counted from 1 to 10 in binary code—effectively announcing to the aliens that numbers will be represented using these symbols.

Having established a way of counting, Drake then moved to connect the concept of numbers to some reference that the citizens of M13 would likely share with us. For this step, he encoded the atomic numbers for five elements: hydrogen, carbon, nitrogen, oxygen, and phosphorous, the building blocks of DNA. Other parts of the message were more visually oriented. Drake used the on-off pulses of the radio signal to "draw" a pixelated image of a human body. He also included a sketch of our solar system and of

the Arecibo telescope itself. The message said, in effect: this is how
we count; this is what we are made of; this is where we came from;
this is what we look like; and this is the technology we are using to
send this message to you.

As inventive as Drake's exosemiotics were in 1974, the Arecibo
message was ultimately more of a proof-of-concept than a genuine
attempt to make contact, as Drake himself is the first to admit. For
starters, the 25,000 light-years that separate us from M13 raise a
legitimate question about whether humans will even be around
—or recognizably human—by the time a message comes back.
The choice of where to send it was almost entirely haphazard. The
METI project intends to improve on the Arecibo model by directly
targeting nearby Goldilocks-zone planets.

One of the most recent planets added to that list orbits the star
Gliese 411, a red dwarf located eight light-years away from Earth.
On a spring evening in the Oakland hills, our own sun putting on
a spectacular display as it slowly set over the Golden Gate Bridge,
Vakoch and I met at one of the observatories at the Chabot Space
and Science Center to take a peek at Gliese 411. A half-moon overhead reduced our visibility but not so much that I couldn't make
out the faint tangerine glimmer of the star, a single blurred point
of light that had traveled nearly 50 trillion miles across the universe to land on my retina. Even with the power of the Oakland
telescope, there was no way to spot a planet orbiting the red dwarf.
But in February of this year, a team of researchers using the Keck I
telescope at the top of Mauna Kea in Hawaii announced that they
had detected a "super-Earth" in orbit around Gliese, a rocky and
hot planet larger than our own.

The METI group aims to improve on the Arecibo message
not just by targeting specific planets, like that super-Earth orbiting Gliese, but also by rethinking the nature of the message itself.
"Drake's original design plays into the bias that vision is universal among intelligent life," Vakoch told me. Visual diagrams—
whether formed through semiprime grids or engraved on plaques
—seem like a compelling way to encode information to us because
humans happen to have evolved an unusually acute sense of vision. But perhaps the aliens followed a different evolutionary path
and found their way to a technologically advanced civilization with

an intelligence that was rooted in some other sense: hearing, for example, or some other way of perceiving the world around them for which there is no earthly equivalent.

Like so much of the SETI/METI debate, the question of visual messaging quickly spirals out into a deeper meditation, in this instance on the connection between intelligence and visual acuity. It is no accident that eyes developed independently so many times over the course of evolution on Earth, given the fact that light conveys information faster than any other conduit. That transmission-speed advantage would presumably apply on other planets in the Goldilocks zone, even if they happened to be on the other side of the Milky Way, and so it seems plausible that intelligent creatures would evolve some sort of visual system as well.

But even more universal than sight would be the experience of time. Hans Freudenthal's *Lincos: Design of a Language for Cosmic Intercourse,* a seminal book of exosemiotics published more than a half century ago, relied heavily on temporal cues in its primer stage. Vakoch and his collaborators have been working with Freudenthal's language in their early drafts for the message. In Lincos, duration is used as a key building block. A pulse that lasts for a certain stretch (say, in human terms, one second) is followed by a sequence of pulses that signify the "word" for one; a pulse that lasts for six seconds is followed by the word for six. The words for basic math properties can be conveyed by combining pulses of different lengths. You might demonstrate the property of addition by sending the words for three and six and then sending a pulse that lasts for nine seconds. "It's a way of being able to point at objects when you don't have anything right in front of you," Vakoch explains.

Other messaging enthusiasts think we needn't bother worrying about primers and common referents. "Forget about sending mathematical relationships, the value of pi, prime numbers or the Fibonacci series," the senior SETI astronomer, Seth Shostak, argued in a 2009 book. "No, if we want to broadcast a message from Earth, I propose that we just feed the Google servers into the transmitter. Send the aliens the World Wide Web. It would take half a year or less to transmit this in the microwave; using infrared lasers shortens the transmit time to no more than two days." Shostak believes that the sheer magnitude of the transmitted data would enable the aliens to decipher it. There is some precedent for this in the history of archeologists studying dead languages: the hardest code to crack is one with only a few fragments.

Sending all of Google would be a logical continuation of Drake's 1974 message, in terms of content if not encoding. "The thing about the Arecibo message is that, in a sense, it's brief but its intent is encyclopedic," Vakoch told me as we waited for the sky to darken in the Oakland hills. "One of the things that we are exploring for our transmission is the opposite extreme. Rather than being encyclopedic, being selective. Instead of this huge digital data dive, trying to do something elegant. Part of that is thinking about what are the most fundamental concepts we need." There is something provocative about the question Vakoch is wrestling with here: of all the many manifestations of our achievements as a species, what's the simplest message we can create that will signal that we're interesting, worthy of an interstellar reply?

But to METI's critics, what he should be worrying about instead is the form that the reply might take: a death ray, or an occupying army.

Before Doug Vakoch had even filed the papers to form the METI nonprofit organization in July 2015, a dozen or so science-and-tech luminaries, including SpaceX's Elon Musk, signed a statement categorically opposing the project, at least without extensive further discussion, on a planetary scale. "Intentionally signaling other civilizations in the Milky Way Galaxy," the statement argued, "raises concerns from all the people of Earth, about both the message and the consequences of contact. A worldwide scientific, political and humanitarian discussion must occur before any message is sent."

One signatory to that statement was the astronomer and science-fiction author David Brin, who has been carrying on a spirited but collegial series of debates with Vakoch over the wisdom of his project. "I just don't think anybody should give our children a fait accompli based on blithe assumptions and assertions that have been untested and not subjected to critical peer review," he told me over a Skype call from his home office in Southern California. "If you are going to do something that is going to change some of the fundamental observable parameters of our solar system, then how about an environmental-impact statement?"

The anti-METI movement is predicated on a grim statistical likelihood: if we do ever manage to make contact with another intelligent life-form, then almost by definition, our new pen pals will be far more advanced than we are. The best way to understand this is to consider, on a percentage basis, just how young our own high-tech civilization actually is. We have been sending structured radio signals from Earth for only the last 100 years. If the universe were exactly 14 billion years old, then it would have taken 13,999,999,900 years for radio communication to be harnessed on our planet. The odds that our message would reach a society that had been tinkering with radio for a shorter, or even similar, period of time would be staggeringly long. Imagine another planet that deviates from our timetable by just a tenth of 1 percent: if they are more advanced than us, then they will have been using radio (and successor technologies) for 14 million years. Of course, depending on where they live in the universe, their signals might take millions of years to reach us. But even if you factor in that transmission lag, if we pick up a signal from another galaxy, we will almost certainly find ourselves in conversation with a more advanced civilization.

It is this asymmetry that has convinced so many future-minded thinkers that METI is a bad idea. The history of colonialism here on Earth weighs particularly heavy on the imaginations of the METI critics. Stephen Hawking, for instance, made this observation in a 2010 documentary series: "If aliens visit us, the outcome would be much as when Columbus landed in America, which didn't turn out well for the Native Americans." David Brin echoes the Hawking critique: "Every single case we know of a more technologically advanced culture contacting a less technologically advanced culture resulted at least in pain."

METI proponents counter the critics with two main arguments. The first is essentially that the horse has already left the barn: given that we have been "leaking" radio waves in the form of *Leave It to Beaver* and the nightly news for decades, and given that other civilizations are likely to be far more advanced than we are, and thus capable of detecting even weak signals, then it seems likely that we are already visible to extraterrestrials. In other words, they know we're here, but they haven't considered us to be worthy of conversation yet. "Maybe in fact there are a lot more civilizations out there, and even nearby planets are populated, but they're simply observing us," Vakoch argues. "It's as if we are in some galactic zoo, and if they've been watching us, it's like watching zebras talking to one another. But what if one of those zebras suddenly turns toward you and with its hooves starts scratching out the prime numbers. You'd relate to that zebra differently!"

Brin thinks that argument dangerously underestimates the difference between a high-power, targeted METI transmission and the passive leakage of media signals, which are far more difficult to detect. "Think about it this way: If you want to communicate with a Boy Scout camp on the other side of the lake, you could kneel down at the end of the lake and slap the water in Morse code," he says. "And if they are spectacularly technologically advanced Boy Scouts who happened also to be looking your way, they might build instruments that would be able to parse out your Morse code. But then you whip out your laser pointer and point it at their dock. That is exactly the order of magnitude difference between picking up [reruns of] *I Love Lucy* from the 1980s, when we were at our noisiest, and what these guys want to do."

METI defenders also argue that the threat of some Klingon-style invasion is implausible, given the distances involved. If, in fact, advanced civilizations were capable of zipping around the galaxy at the speed of light, we would have already encountered them. The much more likely situation is that only communications can travel that fast, and so a malevolent presence on some distant planet will only be able to send us hate mail. But critics think that sense of security is unwarranted. Writing in *Scientific American,* the former chairman of SETI, John Gertz, argued that "a civilization with malign intent that is only modestly more advanced than we are might be able to annihilate Earth with ease by means of a small projectile filled with a self-replicating toxin or nano gray goo; a kinetic missile traveling at an appreciable percentage of the speed of light; or weaponry beyond our imagination."

Brin looks to our own technological progress as a sign of where a more advanced civilization might be in terms of interstellar combat: "It is possible that within just 50 years, we could create an antimatter rocket that could propel a substantial pellet of several kilograms, at half the speed of light at times to intersect with the orbit of a planet within 10 light-years of us." Even a few kilograms colliding at that speed would produce an explosion much greater than the Hiroshima and Nagasaki detonations combined. "And if we could do that in 50 years, imagine what anybody else could do, completely obeying Einstein and the laws of physics."

Interestingly, Frank Drake himself is not a supporter of the METI efforts, though he does not share Hawking's and Musk's fear of interstellar conquistadors. "We send messages all the time, free of charge," he says. "There's a big shell out there now 80 light-years around us. A civilization only a little more advanced than we are can pick those things up. So the point is we are already sending copious amounts of information." Drake believes that any other advanced civilization out there must be doing the same, so scientists like Vakoch should devote themselves to picking up on that chatter instead of trying to talk back. METI will consume resources, Drake says, that would be "better spent listening and not sending."

METI critics, of course, might be right about the frightening sophistication of these other, presumably older civilizations but wrong about the likely nature of their response. Yes, they could be capable of sending projectiles across the galaxy at a quarter of the speed of light. But their longevity would also suggest that they have figured out how to avoid self-destruction on a planetary scale. As Steven Pinker has argued, human beings have become steadily less violent over the last 500 years; per capita deaths from military conflict are most likely at an all-time low. Could this be a recurring pattern throughout the universe, played out on much longer timescales: the older a civilization gets, the less warlike it becomes? In which case, if we do get a message to extraterrestrials, then perhaps they really will come in peace.

These sorts of questions inevitably circle back to the two foundational thought experiments that SETI and METI are predicated upon: the Fermi Paradox and the Drake Equation. The paradox, first formulated by the Italian physicist and Nobel laureate Enrico Fermi, begins with the assumption that the universe contains an unthinkably large number of stars, with a significant percentage of them orbited by planets in the Goldilocks zone. If intelligent life arises on even a small fraction of those planets, then the universe should be teeming with advanced civilizations. And yet to date, we have seen no evidence of those civilizations, even after several decades of scanning the skies through SETI searches. Fermi's question, apparently raised during a lunch conversation at Los Alamos in the early 1950s, was a simple one: "Where is everybody?"

The Drake Equation is an attempt to answer that question. The equation dates back to one of the great academic retreats in the history of scholarship: a 1961 meeting at the Green Bank observatory in West Virginia, which included Frank Drake, a twenty-six-year-old Carl Sagan, and the dolphin researcher (and later psychedelic explorer) John Lilly. During the session, Drake shared his musings on the Fermi Paradox, formulated as an equation. If we start scanning the cosmos for signs of intelligent life, Drake asked, how likely are we to actually detect something? The equation didn't generate a clear answer, because almost all the variables were unknown at the time and continue to be largely unknown a half century later. But the equation had a clarifying effect, nonetheless. In mathematical form, it looks like this:

$$N = R^* \times f_p \times n_e \times f_l \times f_i \times f_c \times L$$

N represents the number of extant, communicative civilizations in the Milky Way. The initial variable R^* corresponds to the rate of star formation in the galaxy, effectively giving you the total number of potential suns that could support life. The remaining variables then serve as a kind of nested sequence of filters: Given the number of stars in the Milky Way, what fraction of those have planets, and how many of those have an environment that can support life? On those potentially hospitable planets, how often does life itself actually emerge, and what fraction of that life evolves into intelligent life, and what fraction of that life eventually leads to a civilization's transmitting detectable signals into space? At the end of his equation, Drake placed the crucial variable L, which is the average length of time during which those civilizations emit those signals.

What makes the Drake Equation so mesmerizing is in part the way it forces the mind to yoke together so many different intellectual disciplines in a single framework. As you move from left to right in the equation, you shift from astrophysics, to the biochemistry of life, to evolutionary theory, to cognitive science, all the way to theories of technological development. Your guess about each value in the Drake Equation winds up revealing a whole worldview: perhaps you think life is rare, but when it does emerge, intelligent life usually follows; or perhaps you think microbial life is ubiquitous throughout the cosmos, but more complex organisms almost never form. The equation is notoriously vulnerable to very different outcomes, depending on the numbers you assign to each variable.

The most provocative value is the last one: L, the average life span of a signal-transmitting civilization. You don't have to be a Pollyanna to defend a relatively high L value. All you need is to believe that it is possible for civilizations to become fundamentally self-sustaining and survive for millions of years. Even if one in a thousand intelligent life-forms in space generates a million-year civilization, the value of L increases meaningfully. But if your L-value is low, that implies a further question: What is keeping it low? Do technological civilizations keep flickering on and off in the Milky Way, like so many fireflies in space? Do they run out of resources? Do they blow themselves up?

Since Drake first sketched out the equation in 1961, two fundamental developments have reshaped our understanding of the problem. First, the numbers on the left-hand side of the equation (representing the amount of stars with habitable planets) have increased by several orders of magnitude. And second, we have been listening for signals for decades and heard nothing. As Brin puts it: "Something is keeping the Drake Equation small. And the difference between all the people in the SETI debates is not whether that's true, but where in the Drake panoply the fault lies."

If the left-hand values keep getting bigger and bigger, the question is which variables on the right-hand side are the filters. As Brin puts it, we want the filter to be behind us, not the one variable, L, that still lies ahead of us. We want the emergence of intelligent life to be astonishingly rare; if the opposite is true, and intelligent life is abundant in the Milky Way, then L values might be low, perhaps measured in centuries and not even millennia. In that case, the adoption of a technologically advanced lifestyle might be effectively simultaneous with extinction. First you invent radio, then you invent technologies capable of destroying all life on your planet, and shortly thereafter you push the button and your civilization goes dark.

The L-value question explains why so many of METI's opponents —like Musk and Hawking—are also concerned with the threat of extinction-level events triggered by other potential threats: super-intelligent computers, runaway nanobots, nuclear weapons, aster-

oids. In a low L-value universe, planet-wide annihilation is an im-
minent possibility. Even if a small fraction of alien civilizations out
there would be inclined to shoot a two-kilogram pellet toward us at
half the speed of light, is it worth sending a message if there's even
the slightest chance that the reply could result in the destruction
of all life on Earth?

Other, more benign explanations for the Fermi Paradox exist.
Drake himself is pessimistic about the L value, but not for dystopian reasons. "It's because we're getting better at technology," he
says. The modern descendants of the TV and radio towers that inadvertently sent Elvis to the stars are far more efficient in terms of
the power they use, which means the "leaked" signals emanating
from Earth are far fainter than they were in the 1950s. In fact, we
increasingly share information via fiber optics and other terrestrial
conduits that have zero leakage outside our atmosphere. Perhaps
technologically advanced societies do flicker on and off like fireflies, but it's not a sign that they're self-destructive; it's just a sign
that they got cable.

But to some METI critics, even a less apocalyptic interpretation of the Fermi Paradox still suggests caution. Perhaps advanced
civilizations tend to reach a point at which they decide, for some
unknown reason, that it is in their collective best interest not to
transmit any detectable signal to their neighbors in the Milky Way.
"That's the other answer for the Fermi Paradox," Vakoch says with
a smile. "There's a Stephen Hawking on every planet, and that's
why we don't hear from them."

In his California home among the redwoods, Frank Drake has a
version of the Arecibo message visually encoded in a very different
format: not a series of radio-wave pulses but as a stained-glass window in his living room. A grid of pixels on a cerulean blue background, it almost resembles a game of Space Invaders. Stained
glass is an appropriate medium, given the nature of the message:
an offering dispatched to unknown beings residing somewhere in
the sky.

There is something about the METI question that forces the
mind to stretch beyond its usual limits. You have to imagine some
radically different form of intelligence, using only your human
intelligence. You have to imagine timescales on which a decision
made in 2017 might trigger momentous consequences 10,000 years

from now. The sheer magnitude of those consequences challenges our usual measures of cause and effect. Whether you believe that the aliens are likely to be warriors or Zen masters, if you think that METI has a reasonable chance of making contact with another intelligent organism somewhere in the Milky Way, then you have to accept that this small group of astronomers and science-fiction authors and billionaire patrons debating semiprime numbers and the ubiquity of visual intelligence may in fact be wrestling with a decision that could prove to be the most transformative one in the history of human civilization.

All of which takes us back to a much more down-to-earth, but no less challenging, question: who gets to decide? After many years of debate, the SETI community established an agreed-upon procedure that scientists and government agencies should follow in the event that the SETI searches actually stumble upon an intelligible signal from space. The protocols specifically ordain that "no response to a signal or other evidence of extraterrestrial intelligence should be sent until appropriate international consultations have taken place." But an equivalent set of guidelines does not yet exist to govern our own interstellar outreach.

One of the most thoughtful participants in the METI debate, Kathryn Denning, an anthropologist at York University in Toronto, has argued that our decisions about extraterrestrial contact are ultimately more political than scientific. "If I had to take a position, I'd say that broad consultation regarding METI is essential, and so I greatly respect the efforts in that direction," Denning says. "But no matter how much consultation there is, it's inevitable that there will be significant disagreement about the advisability of transmitting, and I don't think this is the sort of thing where a simple majority vote or even supermajority should carry the day . . . so this keeps bringing us back to the same key question: is it okay for some people to transmit messages at significant power when other people don't want them to?"

In a sense, the METI debate runs parallel to other existential decisions that we will be confronting in the coming decades, as our technological and scientific powers increase. Should we create superintelligent machines that exceed our own intellectual capabilities by such a wide margin that we cease to understand how their intelligence works? Should we "cure" death, as many technologists are proposing? Like METI, these are potentially among the most momentous decisions human beings will ever make, and yet the number of people actively participating in those decisions —or even aware such decisions are being made—is minuscule.

"I think we need to rethink the message process so that we are sending a series of increasingly inclusive messages," Vakoch says. "Any message that we initially send would be too narrow, too incomplete. But that's okay. Instead, what we should be doing is thinking about how to make the next round of messages better and more inclusive. We ideally want a way to incorporate both technical expertise—people who have been thinking about these issues from a range of different disciplines—and also getting lay input. I think it's often been one or the other. One way we can get lay input in a way that makes a difference in terms of message content is to survey people about what sorts of things they would want to say. It's important to see what the general themes are that people would want to say and then translate those into a Lincos-like message."

When I asked Denning where she stands on the METI issue, she told me: "I have to answer that question with a question: Why are you asking me? Why should my opinion matter more than that of a six-year-old girl in Namibia? We both have exactly the same amount at stake, arguably, she more than I, since the odds of being dead before any consequences of transmission occur are probably a bit higher for me, assuming she has access to clean water and decent healthcare and isn't killed far too young in war." She continued, "I think the METI debate may be one of those rare topics where scientific knowledge is highly relevant to the discussion, but its connection to obvious policy is tenuous at best, because in the final analysis, it's all about how much risk the people of Earth are willing to tolerate . . . And why exactly should astronomers, cosmologists, physicists, anthropologists, psychologists, sociologists, biologists, sci-fi authors, or anyone else (in no particular order) get to decide what those tolerances should be?"

Wrestling with the METI question suggests, to me at least, that the one invention human society needs is more conceptual than technological: we need to define a special class of decisions that potentially create extinction-level risk. New technologies (like superintelligent computers) or interventions (like METI) that pose even the slightest risk of causing human extinction would require some novel form of global oversight. And part of that process would entail establishing, as Denning suggests, some measure of risk tolerance on a planetary level. If we don't, then by default the gamblers will always set the agenda, and the rest of us will have to live with the consequences of their wagers.

In 2017, the idea of global oversight on any issue, however existential the threat it poses, may sound naive. It may also be that technologies have their own inevitability, and we can only rein them in for so long: if contact with aliens is technically possible, then someone, somewhere is going to do it soon enough. There is not a lot of historical precedent for humans voluntarily swearing off a new technological capability—or choosing not to make contact with another society—because of some threat that might not arrive for generations. But maybe it's time that humans learned how to make that kind of choice. This turns out to be one of the surprising gifts of the METI debate, whichever side you happen to take. Thinking hard about what kinds of civilization we might be able to talk to ends up making us think even harder about what kind of civilization we want to be ourselves.

Near the end of my conversation with Frank Drake, I came back to the question of our increasingly quiet planet: all those inefficient radio and television signals giving way to the undetectable transmissions of the internet age. Maybe that's the long-term argument for sending intentional messages, I suggested; even if it fails in our lifetime, we will have created a signal that might enable an interstellar connection thousands of years from now.

Drake leaned forward, nodding. "It raises a very interesting, nonscientific question, which is: Are extraterrestrial civilizations altruistic? Do they recognize this problem and establish a beacon for the benefit of the other folks out there? My answer is: I think it's actually Darwinian; I think evolution favors altruistic societies. So my guess is yes. And that means there might be one powerful signal for each civilization." Given the transit time across the universe, that signal might well outlast us as a species, in which case it might ultimately serve as a memorial as much as a message, like an interstellar version of the Great Pyramids: proof that a technologically advanced organism evolved on this planet, whatever that organism's ultimate fate.

As I stared at Drake's stained-glass Arecibo message, in the middle of that redwood grove, it seemed to me that an altruistic civilization—one that wanted to reach across the cosmos in peace—would be something to aspire to, despite the potential for risk. Do we want to be the sort of civilization that boards up the windows and pretends that no one is home, for fear of some unknown threat lurking in the dark sky? Or do we want to be a beacon?

ELENA PASSARELLO

Arabella (Arancus diadematus) *1973*

FROM *The Normal School*

The tiny insect had, in many ways, been Skylab's star performer.
—Reuters

IN EARLY SUMMER, once she has broken from her cocoon and spent a day or so in the huddle of her family, an adolescent cross spider feels ready to fly. She scuttles to some swatch of vegetation that faces outward—a leaf ridge, a twig—and she perches there. Then she lifts a significant fraction of her legs and pulls a silk strand from the spinnerets at the base of her abdomen. Ounce for ounce, that silk is five times the strength of reinforced steel and at least twice as strong as a human femur.

When a June wind blows the silk strand into its current, she follows it with her body: a tiny balloon chasing its string. It is a journey that begins with a considerable jerk and ends only when her silk tether collides with a rooted object   a bush, a fence post. This could be a one-yard trip, or it could send her half a mile away. A half-mile journey to a cross spider is like a man catching a wind from Milwaukee to Madison.

Gravity left the bodies of the Skylab III crew without much warning. They had spent the early minutes of the launch pressed into their command module's "couch" while 7.7 million pounds of rocket thrust pushed away the Earth. Now, a small space separated their thighs from the seat fabric. If not for the straps of their harnesses, the three men would have risen like milkweed to the low ceilings of the transport pod.

It would be several more hours before a dot appeared in the navigational telescope, unmistakably white against the black of space. More hours still, waiting and hovering, until the dot became a shining oblong cylinder, flanked by two giant solar-paneled flags and topped with a windmill-shaped telescope. Skylab. Here was where they would dock themselves: at the gate of this white and black and brilliant gold capsule, which was falling in the orbit of the planet they'd just escaped at nearly one-half mile per second.

Not too long after landing, a cross spider spins her first full web. She casts a line outward, waits to feel it catch, and then secures the other end of the line to the spot where she rests. This creates a single-strand bridge that she can walk across, which she does, reinforcing it with a second bridge line. Once suspended from the center of that bridge, she free-falls, still pumping silk to form a Y-shape. She then pulls strand after strand from her body, spinning and falling, climbing and plummeting, hooking each strand to the crotch of that Y. Soon, a dozen spokes branch from the Y-hub like a silken sunburst.

Without stopping, she turns sideways and circles the spokes, connecting them in 30 cartwheeled spirals. Here is where she switches the gears of her body to produce a stickier silk—viscid enough to trap heavy prey. With this silk she weaves a second spiral. After that's done, she eats the first spiral, then she eats the hub, and finally she arranges herself in the hub's place. And though she will never rate a vantage to see her handiwork (even if she could, her eyes can't focus at such a distance), the young spider has just filled her space with one of our Earth's most spectacular pieces of craftsmanship, just as versions of herself have done for hundreds of millions of years. It takes her about half an hour.

NASA spent almost a decade designing Skylab's orbital workshop, and its final blueprint held limited consideration for up and down. Rather than separate the station's two levels with a solid floor, a crosshatching of beams split the workshop like an open, metallic net. A long blue pole ran through the center, so the men could pull themselves along the workshop's 48 feet, but the astronauts scrapped the pole shortly after getting their space bearings. They preferred pushing off the walls and steering with their arms, float-

ing through that empty center to travel from the workshop's fore
level—site of the dinner table, the latrine, and the three booths in
which they slept bolted to the walls—to the aft level—with its radio
and TV equipment, its biophysics lab, its materials processors, and
its plastic vial the size of a human thumb containing a young cross
spider named Arabella.

A spider was built to strum her web like a guitar. She was built to
pluck a radial with one striped tarsal claw and feel how the pull
of the world changes the vibration of her web. She was built to
spin more sticky strands at web-bottom than at web-top, as gravity
makes jumping down to prey less taxing than climbing up to it.
She was built to drop a gossamer line and free-fall from danger, to
walk the strands of her handiwork upside down, using her weight
for propulsion.

For nothing says *spider* more than this built-in vigilance, this innate knowledge of what pushes her into the Earth and what lifts
her away from it. Her legs, claws, mouth, the silk she unspools
from inside herself, they all understand—with the hair-trigger sensitivity that comes from eons of experiments—the facts of our massive planet trying to collide with her body.

It wasn't until the eighth day of the mission that Science Pilot
Owen Garriott floated over to Arabella's little vial. NASA had cus-
tom-built her a 15-inch square cage, as narrow as a framed portrait,
with a flat glass front. Around the frame were mounts for long
fluorescent bulbs and cameras, and at the top right corner was an
attachment point for Arabella's transport vial, which would create
a narrow tunnel for her to pass into the cage. Nobody wanted to
risk releasing a tiny spider into the free space of the orbital workshop, which the crew knew had a mind of its own.

She refused to take the tunnel for a full mission day. Though
she was only the responsibility of the science pilot, the two other
crewmen couldn't help but keep tabs on her. Pilot Jack Lousma
watched Garriott opening Arabella's vial and floated over to assist.
"It didn't know where it was, poor spider," he later remembered.
Commander Alan Bean, one of the dozen men who have walked
on the moon, noted her in his private journal: "Owen got the
vial off the cage, opened the door, and shook her out, where she

immediately bounced back and forth, front to back, four or five times, then locked onto the screen panels at the box edge. There she sits, clutching the screen."

The 16-millimeter film of Arabella's earliest Skylab work depicts not so much a spider as the specter of one—a black-and-gray arachno-ghost. Eight thin strands glimmer about her body: these are her legs. White dots sparkle from a dozen other faint pinstripes: these are her gossamer. In the film, she tries to free-fall and hang an early radial, with tumultuous results. You watch her scurry along a horizontal line, half holding on and half bouncing, until she loses all footing; then weightlessness floats her above the line. A flailing of legs sends her tumbling in the other direction, sinking lower, though there is no "lower" to a spider somersaulting in a cage-in-space. She first flips head over abdomen, then corkscrews, so that her rolling turns sideways. Her legs reach outward in eight directions; then they all move inward, clutching the empty space as one desperate claw. Eventually, she finds the hard purchase of the cage's corner and tries to locate some stillness there.

Science Pilot Garriott had spent months rehearsing his upcoming spacewalks on a drowned Skylab mock-up in Huntsville. On Mission Day 9, he shut off the airlock room to the rest of the station, opened the execution hatch, and exited to a place only 25 men had gone before him: out into the thermosphere, the "hard vacuum," the emptiness between the Earth and its closest celestial neighbor. A 60-foot cord spooled out from his abdomen.

Slowed by the fat fingers of his gloves, he was surprised by the effort it took to pull a 5-foot pole of thin steel from the hatch, and then another, then another, linking 11 of them end-to-end until they made two 55-foot-long booms that pointed out into space. He then unhitched himself from the footholds to float a giant sail of thermally treated gold fabric to the attachment points at the poles' ends. Next, he reeled the sail until it covered the orbital workshop like a golden duvet, shielding Skylab from the heat of the sun.

Before returning indoors, Garriott swam to the end of the windmill telescope on the other side of the spacecraft. Since Huntsville, he'd vowed to find a moment in space to put his toes on the edge of Skylab and to look down.

And what he saw was, in a sense, a distance equal to the entire length of the Grand Canyon. In another sense, he saw the equivalent of a a cross spider staring over a small asteroid, perhaps the size of the one named after the schoolteacher who died on her way to space. And in a completely different sense, the science pilot saw nothing at all.

Once back inside Skylab, Garriott discovered Arabella had spun her first web in microgravity.

The distance between a man and the moon is a spider hiking the Oregon Trail. The distance from a spider to the end of her six-inch silk tether is a man drifting on a sixty-foot umbilical. A man tumbling from end to end of a space station is a spider free-falling down a four-foot web.

For a spider is a particle and a man is a particle, and the spider attracts the man with a force directly proportional to the product of their masses and inversely proportional to the square of their distances.

For a man is least distant from a spider when the world he knows is multiplied by cosmic exponents.

For a spider is most distant from a man when she no longer has the tools to refer to herself.

The first web she made was loose and haphazard—a fun-house mirror of her gridded, Earth-spun work. Since she could not feel the weight of her body on the strands, the silk she spun was of varied and impractical thickness. Few of the lines were taut or straight. The web looked like the worn-out shawl of a sideshow palmist, or a sea net from which any fish could manage escape. It is the kind of web only spun on Earth when a spider has just molted, or is quite near her death. It is not unlike a web spun by an amputee spider, or the spider that a Swiss pharmacologist spiked with D-Amphetamine in 1948.

Space was a kind of stimulant for the Skylab crew, who, in the second week of their nine-week flight, radioed Mission Control for extra work. Science Pilot Garriott logged 22 hours in a single mission day. Though NASA had proposed sending them to space with a game console and movie projector, all the crew really wanted to do for R and R was stare out the window and monkey around in the microgravity. Because space awakens things in the human body. The men had become more dexterous and less nauseated. Zero-G decompressed their vertebrae; each man stretched at least an inch by splashdown.

In the evenings, the men floated to the ring of hulking white lockers that encircled one end of the workshop and they ran laps, running the "wheel" from wall to ceiling and back down, like three Astaire hamsters. Or they would tumble and flip in their white socks and boxers, gliding into one another, making water-ballerina shapes. They challenged each other to contests of floating—from the trash airlock, down past Arabella's TV cage, past the kitchen and through the hatches, all the way to the command module that brought them to space—without flapping an appendage.

"We might have had a shot at the Olympics," Garriott later quipped. The science pilot called this feat "playing Spider-Man."

The lights came on in Arabella's cage, simulating sunrise. It was six days after her release from the vial and two weeks after launch, and her air was still charged with that defeating kind of nothing that kept her lines from catching and confounded her body when she tried to rappel. But the walls of the cage were solid in a way she could understand, so she connected a taut bridge line in the short gap between them. The only way to make a web in flight, she discovered, was to avoid flight—to stay grounded.

Clinging to the walls and corners of the cage, she made more short silk lines, pulled them as tightly as she could, and walked along them (rather than free-falling) to affix every radial, every spiral turn. She used the length of her hind legs to measure each spiral; each ring matched the distance between her spinneret and the tip of her back claw. The central claws of her third pair of legs clung to the silk like grappling hooks.

It was a less ambitious web in certain aspects. She had ringed it in fewer spirals and omitted the crucial lopsidedness of lining the web-bottom with that stickier silk. But the structure was even and tight. Back home, a human face crashing through it would have thought it the work of any earthbound arachnid.

*

From Commander Bean's journal, August 7: "Arabella finished her web perfectly. When Owen told Jack at breakfast, Jack said, 'Well, that's good. I like to see a spider do something at least once in a while.'"

Bean again, on August 8: "Arabella ate her web last night, and spun another perfect one."

Commander Bean on August 10: "Owen did the Spider TV three times today."

The men and their cargo were orbiting Earth every other hour, curving 50 degrees north and south of the equator—ringing the parallels, rising and sinking. Their remaining tethers to the Earth were the kind that are difficult to pull on—teleprinter dispatches, the transmitted sinusoidal waves of their wives' voices, the electric grid of Enid, Oklahoma, which blinked for Garriott when his hometown knew he was passing overhead. And though their distance never increased from the day they docked, the human cargo drifted further with each spacewalk, each tube of liquid spaghetti, each wrench slipping from their fingers that they no longer reflexively reached out to catch.

August 11 was supposedly the crew's day off, but none of the three wanted rest. Neither did Arabella, who spun another lovely web— her fourth in as many days. Garriott had taken to calling her "our friend Arabella" and noted the day before that she was "near a very large horizon at this point." In his transmission back to Earth on August 12, he described his friend moving her web from the corners of her cage to front and center—a much more telegenic place—where it stretched ever closer to the webs of home. There was a delighted lilt in his voice when he reported how, "without the benefit of previous experience and simply working on her own, she figured out a very nice solution to the problems of zero gravity."

Houston woke the men every "morning" with news updates, but the crew knew the worst events of the day were always missing. NASA had arranged for Skylab to hear of no plane crash at Logan Airport, no coup in Chile, no serial killer found 15 miles from the very control room that shot their news into the sky. Back on Earth, Skylab was, of course, making its own headlines—at least an article a day on the men and their work hounding comets and setting records for nights slept in orbit. And most of the stories found a way to weave into their paragraphs news of the little spider in the TV-ready box.

Footage of her spinning aired on the *CBS Evening News* with Walter Cronkite. Her image, perched at the center of an impressive silver-toned web, ran in *Science News*, the *Los Angeles Times*, and along the AP wire. With each web spun, observed, photographed, and transmitted, things became stickier between Arabella, the astronauts, and the people of Earth. NASA had planned to let her die in space after a few days, but as the *Washington Post* said, Arabella had "earned the affection of the crew" so completely that, "midway through the flight, Owen Garriott asked what could be done to prolong that relationship."

Who could possibly condemn to death the weaver of the universe's first space web? A creature that journalist Lee Edson called "probably the most distinguished spider in the world"? And how could anyone be surprised when Edson announced, "NASA has made another high-level decision. Arabella will be permitted to return to Earth with the astronauts on September 25."

By then, Mission Control had its own mascot spider, Arachne, living in a glass cage in the Command Center. They could look at her when they heard Garriott in space, describing his friend Arabella. They radioed an order for Garriott to set aside a housefly-sized morsel of that evening's space-dinner: filet mignon. "When Arabella is in her cage, carefully place a piece near her legs," ordered CAPCOM Richard Truly. According to the *New York Times*, "The spider loved it, and proceeded to build another web."

Each new web held the planet tighter in her grip. Only Arabella could bridge the distance between all earthbound creatures and the incomprehensible developments of Skylab—the X-ray maps of the galaxy, the Technicolor evidence of holes in the sun. As she steadied her slow steps along the straight line of her web and gripped the solid products of her body for support, she was walking back to what the people of Earth could understand: an organic narrative of success amid what on TV and in the papers must have seemed a dark and unnatural nebula. She did so naked, assisted by nothing but the tangible: her legs, her cage, the light bulb someone turned on for her every 18 hours. Dogged and stealthy, the spider created an earthly object in that wild and distant nothing. And alongside all the work of the floating men, her work was weighty and familiar—a lifeline tailor-made for the people of Earth.

For a spider in the center of her web is less distant from you than a man backflipping through a spaceship in his underwear. For a man in space is a decorated navy pilot in alien coveralls, unraveling a Golden Fleece to save his billion-dollar ship from frying in the nearby sun. For the spider in space still only knows a *garden* sun. She lifts the same eight unsheathed legs that tread any apple branch. She isn't an honorary Doctor of Science or a Fellow at the American Astronomical Society. She's never seen the Earth while standing on its moon.

For a spider in space has no title, just the sweet name of one of our daughters. When we speak it, the name makes the sound of a bell in the air.

Commander Alan Bean retired a few years after splashdown to start a painting career, and nearly every canvas he finishes is a scene in oil of a man on the moon. After commanding a space shuttle voyage in the early 1980s, Pilot Jack Lousma ran for state senator and lost. Owen Garriott returned to space a decade after Skylab, then helped send his son to the International Space Station 25 years later—the first American to pass space travel down a generation.

Arabella made it back to Earth, but just barely. According to NASA, they found the spider curled into a ball in her transport vial the day after splashdown. "An autopsy will be performed," reassured Reuters. It ruled dehydration as the cause of her death, and then there was nothing else to do but catalogue her with the rest of the Skylab data.

Preserved in formalin and arranged in a black cylinder with a Plexiglas viewing front, she's now item A19740484001 in the National Air and Space Museum's collection, occasionally on display with Skylab's other equipment. Among the titanium alloy, the neoprene, the epoxy-resin ablative, and the resin-impregnated fiberglass honeycomb, she is the rare logged item listed as "organic matter."

Though he's outlived her by over 40 years, Garriott has not forgotten the spider. In 2013, he sat on a NASA panel held for the 40th anniversary of Skylab—a panel that also included a science pilot from the contemporary space station. When this younger scientist referenced early work with animals in microgravity, Garriott, now a vigorous eighty-two-year-old, all but interrupted him. He sat forward in his chair and pushed his words out faster than he could pronounce them. His blue eyes widened as he looked into the small crowd of reporters and students. "Does the name Arabella ring a bell with any of you?"

Behind him was a large projection of the official Skylab III mission patch, which had been sewn to the shoulders of all Skylab personnel. The circular patch is a take on Leonardo da Vinci's *Vitruvian Man*, in his notorious power stance. Behind the naked man is a circle half filled with a globe and half filled with a flaming sun. The man seems to float in the center of the patch, splitting the blue and the orange. He reaches to the edges of the frame with all eight of his appendages.

ED YONG

Tiny Jumping Spiders Can See the Moon

FROM *The Atlantic*

LAST WEDNESDAY, A SPIDER fell onto Jamie Lomax's laptop. Two days later, it happened again. Soon enough, several spiders were crawling across the ceiling of her office. "It was a little unnerving," says Lomax, who's an astronomer at the University of Washington. "I'm not scared of spiders but if someone else wants to take care of the spider in a room, I'll gladly let them do it over me. And I don't really want them raining down on my head."

Lomax identified the abseiling arachnids as zebra jumping spiders, and tweeted about her experiences with the hashtag #ItIsRainingSpidersNotMen. And after considering options including "nukes and fire," she settled for notifying her university. They sent over an exterminator, who failed to find any lingering spiders within the ceiling. He figured that a nest had probably hatched, and the newborn spiders had scattered. "But a couple of hours later, there were still spiders everywhere," she tells me. "As of yesterday, there still were."

Meanwhile, fellow astronomer Alex Parker had read Lomax's tweets. "Have you tried lasers?" he replied. "Seriously though, some jumping spiders will chase laser pointers like cats do."

There are, indeed, many YouTube videos of them doing exactly that. But Emily Levesque—Lomax's colleague, with an office two doors down—wanted to see it for herself. "She has a laser pointer and she happens to be the only other person with spiders in her office," says Lomax. "She ran down to me and said: You have to see this."

Being scientists, Lomax and Levesque tested laser pointers of different colors. They found that the zebra spiders seemed to be mildly interested in a red dot but completely transfixed by a green one. They even tried using both lasers at the same time—and the spiders seemed to prefer green over red.

"Do all zebra spiders react more to green vs red laser pointers?" Levesque tweeted. "We need some kind of 'science Twitter' bat signal that we can put up when different fields need input from one another."

But on Twitter, no such signal is necessary. Last night, spider researcher Catherine Scott saw the thread and looped in her friend Nate Morehouse, who studies spider vision at the University of Cincinnati. Morehouse was up late watching the Stanley Cup final and was distraught to see his team, the Pittsburgh Penguins, losing to the Nashville Predators. "I was all bummed out, and I decided to check Twitter before I went to bed," he says. "I had like 150 notifications."

"We can explain all of this!" he wrote to Levesque. Jumping spiders are visual hunters, which track their prey with the large pair of eyes at the front of their heads. The retinas of those eyes contain two types of light-detecting cells—one that's sensitive to ultraviolet light and another that's sensitive to green light. The latter cells aren't *only* sensitive to green light; they react to red too, just less strongly. So jumping spiders can see red light, but it would just appear as a dimmer form of green to them.

No one has really studied the eyes of the zebra jumping spider, says Morehouse, but based on what we know from other species, it should react to laser pointers in exactly the way that Levesque and Lomax found. "It's not a controlled experiment, and the green laser might just be brighter or larger than the red one," Morehouse tells me. "But if you had two equal laser pointers, one red and one green, we'd expect that the jumping spider should track the red one less enthusiastically."

But there's another reason why jumping spiders should fascinate astronomers, besides their occasional penchant for raining from the ceiling or chasing lasers. As Morehouse told Lomax and Levesque, their eyes "are built like . . . wait for it . . . Galilean telescopes." These telescopes, which Galileo started using in 1609, are basically tubes with a lens at each end. Only three groups of animals have similar eyes: falcons, chameleons, and jumping spiders.

In the spider's case, each of the two main eyes is topped with a large lens that's fixed to the rest of the spider's body. Beneath that is a long tube, filled with a clear gel. And at the bottom of the tube, the gel changes in a way that we still don't understand, but that causes light to bend. It effectively acts like a second lens, even though there's no distinct physical structure that you can dissect out.

The two lenses work in tandem: the top one collects and focuses light, while the bottom one spreads it out. This arrangement enlarges images before they hit the spider's retina, which allows it to resolve a huge amount of detail for its size. A jumping spider can see objects as clearly as a pigeon or a small dog, even though its eye tube is less than a millimeter long, and its whole *body* gets no bigger than five millimeters.

Scientists know all of this because they can peer straight into a jumping spider's eyes and study the retinas below. Those retinas have muscles and can swivel around like the back of a telescope, so the spider can change where it's looking without moving its head. By watching them do this, and measuring their anatomy, people like Morehouse can work out how light travels through their eyes. And by extension, they can also calculate what sorts of things the spiders can see.

And as Morehouse told Lomax, Levesque, and the other gathered astronomers on Twitter, "They can definitely resolve the moon in the night sky."

The astronomers did some quick calculations and worked out that the spiders can indeed see the moon, but not planets like Jupiter or Mars. They have enough spatial acuity for seeing Andromeda—our nearest galaxy—but they probably can't make out light that dim. "The unfortunate thing about this crazy telescope-like eye is that it's not as good for capturing light as something designed for night vision," says Morehouse.

The moon, however, is almost certainly bright enough for them. Its light would hit between five and six light-detecting cells on their retinas, which might just be enough for them to make out the subtle darkness of the moon's craters. "They could, in principle, see variation across that surface," says Morehouse. "I'd have to look at luminance differences across the moon, but it's possible."

There's something rather magical about all of this. Twitter has been variously described as an echo chamber where people go to have their views confirmed, or a cesspit where harassment and abuse go unchecked. But occasionally, it is also a place where astronomers with a sudden interest in spider vision can meet spider scientists with a sudden interest in the stars, and geek out over laser-chasing arachnids with telescopes for eyes.

PART VII

"Stop Inhibiting My Action Potential"

Neuroscience and Psychology

BARBARA BRADLEY HAGERTY

When Your Child Is a Psychopath

FROM *The Atlantic*

THIS IS A good day, Samantha tells me: 10 on a scale of 10. We're sitting in a conference room at the San Marcos Treatment Center, just south of Austin, Texas, a space that has witnessed countless difficult conversations between troubled children, their worried parents, and clinical therapists. But today promises unalloyed joy. Samantha's mother is visiting from Idaho, as she does every six weeks, which means lunch off campus and an excursion to Target. The girl needs supplies: new jeans, yoga pants, nail polish.

At eleven, Samantha is just over five feet tall and has wavy black hair and a steady gaze. She flashes a smile when I ask about her favorite subject (history), and grimaces when I ask about her least favorite (math). She seems poised and cheerful, a normal preteen. But when we steer into uncomfortable territory—the events that led her to this juvenile-treatment facility nearly 2,000 miles from her family—Samantha hesitates and looks down at her hands. "I wanted the whole world to myself," she says. "So I made a whole entire book about how to hurt people."

Starting at age six, Samantha began drawing pictures of murder weapons: a knife, a bow and arrow, chemicals for poisoning, a plastic bag for suffocating. She tells me that she pretended to kill her stuffed animals.

"You were practicing on your stuffed animals?" I ask her.

She nods.

"How did you feel when you were doing that to your stuffed animals?"

"Happy."

"Why did it make you feel happy?"

"Because I thought that someday I was going to end up doing it on somebody."

"Did you ever try?"

Silence.

"I choked my little brother."

Samantha's parents, Jen and Danny, adopted Samantha when she was two. They already had three biological children, but they felt called to add Samantha (not her real name) and her half sister, who is two years older, to their family. They later had two more kids.

From the start, Samantha seemed a willful child, in tyrannical need of attention. But what toddler isn't? Her biological mother had been forced to give her up because she'd lost her job and home and couldn't provide for her four children, but there was no evidence of abuse. According to documentation from the state of Texas, Samantha met all her cognitive, emotional, and physical milestones. She had no learning disabilities, no emotional scars, no signs of ADHD or autism.

But even at a very young age, Samantha had a mean streak. When she was about twenty months old, living with foster parents in Texas, she clashed with a boy in daycare. The caretaker soothed them both; problem solved. Later that day Samantha, who was already potty trained, walked over to where the boy was playing, pulled down her pants, and peed on him. "She knew exactly what she was doing," Jen says. "There was an ability to wait until an opportune moment to exact her revenge on someone."

When Samantha got a little older, she would pinch, trip, or push her siblings and smile if they cried. She would break into her sister's piggy bank and rip up all the bills. Once, when Samantha was five, Jen scolded her for being mean to one of her siblings. Samantha walked upstairs to her parents' bathroom and washed her mother's contact lenses down the drain. "Her behavior wasn't impulsive," Jen says. "It was very thoughtful, premeditated."

Jen, a former elementary school teacher, and Danny, a physician, realized they were out of their depth. They consulted doctors, psychiatrists, and therapists. But Samantha only grew more dangerous. They had her admitted to a psychiatric hospital three times before sending her to a residential treatment program in Montana at age six. Samantha would grow out of it, one psychologist assured her parents; the problem was merely delayed empathy. Samantha was impulsive, another said, something that medication would fix. Yet another suggested that she had reactive attachment disorder, which could be ameliorated with intensive therapy. More darkly—and typically, in these sorts of cases—another psychologist blamed Jen and Danny, implying that Samantha was reacting to harsh and unloving parenting.

One bitter December day in 2011, Jen was driving the children along a winding road near their home. Samantha had just turned six. Suddenly Jen heard screaming from the back seat, and when she looked in the mirror, she saw Samantha with her hands around the throat of her two-year-old sister, who was trapped in her car seat. Jen separated them, and once they were home, she pulled Samantha aside.

"What were you doing?," Jen asked.

"I was trying to choke her," Samantha said.

"You realize that would have killed her? She would not have been able to breathe. *She would have died.*"

"I know."

"What about the rest of us?"

"I want to kill all of you."

Samantha later showed Jen her sketches, and Jen watched in horror as her daughter demonstrated how to strangle or suffocate her stuffed animals. "I was so terrified," Jen says. "I felt like I had lost control."

Four months later, Samantha tried to strangle her baby brother, who was just two months old.

Jen and Danny had to admit that nothing seemed to make a difference—not affection, not discipline, not therapy. "I was reading and reading and reading, trying to figure out what diagnosis made sense," Jen tells me. "What fits with the behaviors I'm seeing?" Eventually she found one condition that did seem to fit—but it was a diagnosis that all the mental-health professionals had dismissed, because it's considered both rare and untreatable. In July 2013, Jen took Samantha to see a psychiatrist in New York City, who confirmed her suspicion.

"In the children's mental-health world, it's pretty much a terminal diagnosis, except your child's not going to die," Jen says. "It's just that there's no help." She recalls walking out of the psychiatrist's office on that warm afternoon and standing on a street corner in Manhattan as pedestrians pushed past her in a blur. A feeling flooded over her, singular, unexpected. Hope. Someone had finally acknowledged her family's plight. Perhaps she and Danny could, against the odds, find a way to help their daughter.

Samantha was diagnosed with conduct disorder with callous and unemotional traits. She had all the characteristics of a budding psychopath.

Psychopaths have always been with us. Indeed, certain psychopathic traits have survived because they're useful in small doses: the cool dispassion of a surgeon, the tunnel vision of an Olympic athlete, the ambitious narcissism of many a politician. But when these attributes exist in the wrong combination or in extreme forms, they can produce a dangerously antisocial individual, or even a cold-blooded killer. Only in the past quarter century have researchers zeroed in on the early signs that indicate a child could be the next Ted Bundy.

Researchers shy away from calling children psychopaths; the term carries too much stigma, and too much determinism. They prefer to describe children like Samantha as having "callous and unemotional traits," shorthand for a cluster of characteristics and behaviors, including a lack of empathy, remorse, or guilt; shallow emotions; aggression and even cruelty; and a seeming indifference to punishment. Callous and unemotional children have no trouble hurting others to get what they want. If they do seem caring or empathetic, they're probably trying to manipulate you.

Researchers believe that nearly 1 percent of children exhibit these traits, about as many as have autism or bipolar disorder. Until recently, the condition was seldom mentioned. Only in 2013 did the American Psychiatric Association include callous and unemotional traits in its diagnostic manual, *DSM-5*. The condition can go unnoticed because many children with these traits—who can be charming and smart enough to mimic social cues—are able to mask them.

More than 50 studies have found that kids with callous and unemotional traits are more likely than other kids (three times more likely, in one study) to become criminals or display aggressive, psychopathic traits later in life. And while adult psychopaths constitute only a tiny fraction of the general population, studies suggest that they commit half of all violent crimes. Ignore the problem, says Adrian Raine, a psychologist at the University of Pennsylvania, "and it could be argued we have blood on our hands."

Researchers believe that two paths can lead to psychopathy: one dominated by nature, the other by nurture. For some children, their environment—growing up in poverty, living with abusive parents, fending for themselves in dangerous neighborhoods—can turn them violent and coldhearted. These kids aren't born callous and unemotional; many experts suggest that if they're given a reprieve from their environment, they can be pulled back from psychopathy's edge.

But other children display callous and unemotional traits even though they are raised by loving parents in safe neighborhoods. Large studies in the United Kingdom and elsewhere have found that this early onset condition is highly hereditary, hardwired in the brain—and especially difficult to treat. "We'd like to think a mother and father's love can turn everything around," Raine says. "But there are times where parents are doing the very best they can, but the kid—even from the get-go—is just a bad kid."

Still, researchers stress that a callous child—even one who was born that way—is not automatically destined for psychopathy. By some estimates, four out of five children with these traits do not grow up to be psychopaths. The mystery—the one everyone is trying to solve—is why some of these children develop into normal adults while others end up on death row.

A trained eye can spot a callous and unemotional child by age three or four. Whereas normally developing children at that age grow agitated when they see other children cry—and either try to comfort them or bolt the scene—these kids show a chilly detachment. In fact, psychologists may even be able to trace these traits back to infancy. Researchers at King's College London tested more than 200 five-week-old babies, tracking whether they preferred looking at a person's face or at a red ball. Those who favored the ball displayed more callous traits two and a half years later.

As a child gets older, more-obvious warning signs appear. Kent Kiehl, a psychologist at the University of New Mexico and the author of *The Psychopath Whisperer*, says that one scary harbinger occurs when a kid who is eight, nine, or ten years old commits a transgression or a crime while alone, without the pressure of peers. This reflects an interior impulse toward harm. Criminal versatility—committing different types of crimes in different settings —can also hint at future psychopathy.

But the biggest red flag is early violence. "Most of the psychopaths I meet in prison had been in fights with teachers in elementary school or junior high," Kiehl says. "When I'd interview them, I'd say, 'What's the worst thing you did in school?' And they'd say, 'I beat the teacher unconscious.' You're like, *That really happened?* It turns out that's very common."

We have a fairly good idea of what an adult psychopathic brain looks like, thanks in part to Kiehl's work. He has scanned the brains of hundreds of inmates at maximum-security prisons and chronicled the neural differences between average violent convicts and psychopaths. Broadly speaking, Kiehl and others believe that the psychopathic brain has at least two neural abnormalities—and that these same differences likely also occur in the brains of callous children.

The first abnormality appears in the limbic system, the set of brain structures involved in, among other things, processing emotions. In a psychopath's brain, this area contains less gray matter. "It's like a weaker muscle," Kiehl says. A psychopath may understand, intellectually, that what he is doing is wrong, but he doesn't *feel* it. "Psychopaths know the words but not the music" is how Kiehl describes it. "They just don't have the same circuitry."

In particular, experts point to the amygdala—a part of the limbic system—as a physiological culprit for coldhearted or violent behavior. Someone with an undersized or underactive amygdala may not be able to feel empathy or refrain from violence. For example, many psychopathic adults and callous children do not recognize fear or distress in other people's faces. Essi Viding, a professor of developmental psychopathology at University College London, recalls showing one psychopathic prisoner a series of faces with different expressions. When the prisoner came to a fearful face, he said, "I don't know what you call this emotion, but it's what people look like just before you stab them."

Why does this neural quirk matter? Abigail Marsh, a researcher at Georgetown University who has studied the brains of callous and unemotional children, says that distress cues, such as fearful or sad expressions, signal submission and conciliation. "They're designed to prevent attacks by raising the white flag. And so if you're not sensitive to these cues, you're much more likely to attack somebody whom other people would refrain from attacking."

Psychopaths not only fail to recognize distress in others; they may not feel it themselves. The best physiological indicator of which young people will become violent criminals as adults is a low resting heart rate, says Adrian Raine of the University of Pennsylvania. Longitudinal studies that followed thousands of men in Sweden, the United Kingdom, and Brazil all point to this biological anomaly. "We think that low heart rate reflects a lack of fear, and a lack of fear could predispose someone to committing fearless criminal-violence acts," Raine says. Or perhaps there is an "optimal level of physiological arousal," and psychopathic people seek out stimulation to increase their heart rate to normal. "For some kids, one way of getting this arousal jag in life is by shoplifting, or joining a gang, or robbing a store, or getting into a fight." Indeed, when Daniel Waschbusch, a clinical psychologist at Penn State Hershey Medical Center, gave the most severely callous and unemotional children he worked with a stimulative medication, their behavior improved.

The second hallmark of a psychopathic brain is an overactive reward system especially primed for drugs, sex, or anything else that delivers a ping of excitement. In one study, children played a computer gambling game programmed to allow them to win early on and then slowly begin to lose. Most people will cut their losses at some point, Kent Kiehl notes, "whereas the psychopathic, callous unemotional kids keep going until they lose everything." Their brakes don't work, he says.

Faulty brakes may help explain why psychopaths commit brutal crimes: their brains ignore cues about danger or punishment. "There are all these decisions we make based on threat, or the fear that something bad can happen," says Dustin Pardini, a clinical psychologist and an associate professor of criminology at Arizona State University. "If you have less concern about the negative consequences of your actions, then you'll be more likely to continue engaging in these behaviors. And when you get caught, you'll be less likely to learn from your mistakes."

Researchers see this insensitivity to punishment even in some toddlers. "These are the kids that are completely unperturbed by the fact that they've been put in time-out," says Eva Kimonis, who works with callous children and their families at the University of New South Wales in Australia. "So it's not surprising that they keep going to time-out, because it's not effective for them. Whereas reward—they're very motivated by that."

This insight is driving a new wave of treatment. What's a clinician to do if the emotional, empathetic part of a child's brain is broken but the reward part of the brain is humming along? "You co-opt the system," Kiehl says. "You work with what's left."

With each passing year, both nature and nurture conspire to steer a callous child toward psychopathy and block his exits to a normal life. His brain becomes a little less malleable; his environment grows less forgiving as his exhausted parents reach their limits, and as teachers, social workers, and judges begin to turn away. By his teenage years, he may not be a lost cause, since the rational part of his brain is still under construction. But he can be one scary dude.

Like the guy standing 20 feet away from me in the North Hall of Mendota Juvenile Treatment Center, in Madison, Wisconsin. The tall, lanky teenager has just emerged from his cell. Two staff members cuff his wrists, shackle his feet, and begin to lead him away. Suddenly he swivels to face me and laughs—a menacing laugh that gives me chills. As young men yell expletives, banging on the metal doors of their cells, and others stare silently through their narrow Plexiglas windows, I think, *This is as close as I get to* Lord of the Flies.

The psychologists Michael Caldwell and Greg Van Rybroek thought much the same thing when they opened the Mendota facility in 1995, in response to a nationwide epidemic of youth violence in the early 1990s. Instead of placing young offenders in a juvenile prison until they were released to commit more—and more violent—crimes as adults, the Wisconsin legislature set up a new treatment center to try to break the cycle of pathology. Mendota would operate within the Department of Health Services, not the Department of Corrections. It would be run by psychologists and psychiatric-care technicians, not wardens and guards. It would employ one staff member for every three kids—quadruple the ratio at other juvenile-corrections facilities.

Caldwell and Van Rybroek tell me that the state's high-security juvenile-corrections facility was supposed to send over its most mentally ill boys between the ages of twelve and seventeen. It did, but what Caldwell and Van Rybroek didn't anticipate was that the

boys the facility transferred were also its most menacing and re-
calcitrant. They recall their first few assessments. "The kid would
walk out, and we would turn to each other and say, 'That's the
most dangerous person I've ever seen in my life,'" Caldwell says.
Each one seemed more threatening than the last. "We're looking
at each other and saying, 'Oh, no. What have we done?'" Van Ry-
brock adds.

What they have done, by trial and error, is achieve something
most people thought impossible: if they haven't *cured* psychopathy,
they've at least tamed it.

Many of the teenagers at Mendota grew up on the streets, without parents, and were beaten up or sexually abused. Violence became a defense mechanism. Caldwell and Van Rybroek recall a
group-therapy session a few years ago in which one boy described
being strung up by his wrists and hung from the ceiling as his
father cut him with a knife and rubbed pepper in the wounds.
"Hey," several other kids said, "that's like what happened to me."
They called themselves the "piñata club."

But not everyone at Mendota was "born in hell," as Van Rybroek
puts it. Some of the boys were raised in middle-class homes with
parents whose major sin was not abuse but paralysis in the face of
their terrifying child. No matter the history, one secret to divert-
ing them from adult psychopathy is to wage an unrelenting war
of presence. At Mendota, the staff calls this "decompression." The
idea is to allow a young man who has been living in a state of chaos
to slowly rise to the surface and acclimate to the world without
resorting to violence.

Caldwell mentions that, two weeks ago, one patient became furious over some perceived slight or injustice; every time the techs
checked on him, he would squirt urine or feces through the door.
(This is a popular pastime at Mendota.) The techs would dodge it
and return 20 minutes later, and he would do it again. "This went
on for several days," Caldwell says. "But part of the concept of decompression is that the kid's going to get tired at some point. And
one of those times you're going to come there and he's going to
be tired, or he's just not going to have any urine left to throw at
you. And you're going to have a little moment where you're going
to have a positive connection there."

Cindy Ebsen, the operations director, who is also a registered
nurse, gives me a tour of Mendota's North Hall. As we pass the

metal doors with their narrow windows, the boys peer out and the yelling subsides into entreaties. "Cindy, Cindy, can you get me some candy?" "I'm your favorite, aren't I, Cindy?" "Cindy, why don't you visit me anymore?"

She pauses to banter with each of them. The young men who pass through these halls have murdered and maimed, carjacked and robbed at gunpoint. "But they're still kids. I love working with them, because I see the most success in this population," as opposed to older offenders, Ebsen says. For many, friendship with her or another staff member is the first safe connection they've known.

Forming attachments with callous kids is important, but it's not Mendota's singular insight. The center's real breakthrough involves deploying the anomalies of the psychopathic brain to one's advantage—specifically, downplaying punishment and dangling rewards. These boys have been expelled from school, placed in group homes, arrested, and jailed. If punishment were going to rein them in, it would have by now. But their brains do respond, enthusiastically, to rewards. At Mendota, the boys can accumulate points to join ever more prestigious "clubs" (Club 19, Club 23, the VIP Club). As they ascend in status, they earn privileges and treats —candy bars, baseball cards, pizza on Saturdays, the chance to play Xbox or stay up late. Hitting someone, throwing urine, or cussing out the staff costs a boy points—but not for long, since callous and unemotional kids aren't generally deterred by punishment.

I am, frankly, skeptical—will a kid who knocked down an elderly lady and stole her Social Security check (as one Mendota resident did) really be motivated by the promise of Pokémon cards? But then I walk down the South Hall with Ebsen. She stops and turns toward a door on our left. "Hey," she calls, "do I hear internet radio?"

"Yeah, yeah, I'm in the VIP Club," a voice says. "Can I show you my basketball cards?"

Ebsen unlocks the door to reveal a skinny seventeen-year-old boy with a nascent mustache. He fans out his collection. "This is, like, 50 basketball cards," he says, and I can almost see his reward centers glowing. "I have the most and best basketball cards here." Later, he sketches out his history for me: His stepmother had routinely beat him and his stepbrother had used him for sex. When he was still a preteen, he began molesting the younger girl and boy next door. The abuse continued for a few years, until the boy told his mother. "I knew it was wrong, but I didn't care," he says. "I just wanted the pleasure."

At Mendota, he has begun to see that short-term pleasure could land him in prison as a sex offender, while deferred gratification can confer more-lasting dividends: a family, a job, and most of all, freedom. Unlikely as it sounds, this revelation sprang from his ardent pursuit of basketball cards.

After he details the center's point system (a higher math that I cannot follow), the boy tells me that a similar approach should translate into success in the outside world—as if the world, too, operates on a point system. Just as consistent good behavior confers basketball cards and internet radio inside these walls, so—he believes—will it bring promotions at work. "Say you're a cook; you can [become] a waitress if you're doing really good," he says. "That's the way I look at it."

He peers at me, as if searching for confirmation. I nod, hoping that the world will work this way for him. Even more, I hope his insight will endure.

In fact, the program at Mendota has changed the trajectory for many young men, at least in the short term. Caldwell and Van Rybroek have tracked the public records of 248 juvenile delinquents after their release. One hundred forty-seven of them had been in a juvenile-corrections facility, and 101 of them—the harder, more psychopathic cases—had received treatment at Mendota. In the four and a half years since their release, the Mendota boys have been far less likely to reoffend (64 percent versus 97 percent), and far less likely to commit a violent crime (36 percent versus 60 percent). Most striking, the ordinary delinquents have killed 16 people since their release. The boys from Mendota? Not one.

"We thought that as soon as they walked out the door, they'd last maybe a week or two and they'd have another felony on their record," Caldwell says. "And when the data first came back that showed that that wasn't happening, we figured there was something wrong with the data." For two years, they tried to find mistakes or alternative explanations, but eventually they concluded that the results were real.

The question they are trying to answer now is this: can Mendota's treatment program not only change the behavior of these teens but measurably reshape their brains as well? Researchers are optimistic, in part because the decision-making part of the brain continues to evolve into one's mid-twenties. The program is like neural weight lifting, Kent Kiehl, at the University of New Mexico, says. "If you exercise this limbic-related circuitry, it's going to get better."

To test this hypothesis, Kiehl and the staff at Mendota are now asking some 300 young men to slide into a mobile brain scanner. The scanner records the shape and size of key areas of the boys' brains, as well as how their brains react to tests of decision-making ability, impulsivity, and other qualities that go to the core of psychopathy. Each boy's brain will be scanned before, during, and at the end of their time in the program, offering researchers insights into whether his improved behavior reflects better functioning inside his brain.

No one believes that Mendota graduates will develop true empathy or a heartfelt moral conscience. "They may not go from the Joker in *The Dark Knight* to Mister Rogers," Caldwell tells me, laughing. But they can develop a *cognitive* moral conscience, an intellectual awareness that life will be more rewarding if they play by the rules. "We're just happy if they stay on this side of the law," Van Rybroek says. "In our world, that's huge."

How many can stay the course for a lifetime? Caldwell and Van Rybroek have no idea. They're barred from contacting former patients—a policy meant to ensure that the staff and former patients maintain appropriate boundaries. But sometimes graduates write or call to share their progress, and among these correspondents, Carl, now thirty-seven, stands out.

Carl (not his real name) emailed a thankful note to Van Rybroek in 2013. Aside from one assault conviction after he left Mendota, he had stayed out of trouble for a decade and opened his own business—a funeral home near Los Angeles. His success was especially significant because he was one of the harder cases, a boy from a good home who seemed wired for violence.

Carl was born in a small town in Wisconsin. The middle child of a computer programmer and a special-education teacher, "he came out angry," his father recalls during a phone conversation. His acts of violence started small—hitting a classmate in kindergarten—but quickly escalated: ripping the head off his favorite teddy bear, slashing the tires on the family car, starting fires, killing his sister's hamster.

His sister remembers Carl, when he was about eight, swinging their cat in circles by its tail, faster and faster, and then letting go. "And you hear her hit the wall." Carl just laughed.

Looking back, even Carl is puzzled by the rage that coursed through him as a child. "I remember when I bit my mom really hard, and she was bleeding and crying. I remember feeling so happy, so overjoyed—completely fulfilled and satisfied," he tells me on the phone. "It wasn't like someone kicked me in the face and I was trying to get him back. It was more like a weird, hard-to-explain feeling of hatred."

His behavior confused and eventually terrified his parents. "It just got worse and worse as he got bigger," his father tells me. "Later, when he was a teenager and occasionally incarcerated, I was happy about it. We knew where he was and that he'd be safe, and that took a load off the mind."

By the time Carl arrived at Mendota Juvenile Treatment Center in November 1995, at age fifteen, he had been placed in a psychiatric hospital, a group home, foster care, or a juvenile-corrections center about a dozen times. His police record listed 18 charges, including armed burglary and 3 "crimes against persons," one of which sent the victim to the hospital. Lincoln Hills, a high-security juvenile-corrections facility, foisted him on Mendota after he accumulated more than 100 serious infractions in less than four months. On an assessment called the Youth Psychopathy Checklist, he scored 38 out of a possible 40—five points higher than the average for Mendota boys, who were among the most dangerous young men in Wisconsin.

Carl had a rocky start at Mendota: weeks of abusing staff, smearing feces around his cell, yelling all night, refusing to shower, and spending much of the time locked in his room, not allowed to mix with the other kids. Slowly, though, his psychology began to shift. The staff's unruffled constancy chipped away at his defenses. "These people were like zombies," Carl recalls, laughing. "You could punch them in the face and they wouldn't do anything."

He started talking in therapy and in class. He quit mouthing off and settled down. He developed the first real bonds in his young life. "The teachers, the nurses, the staff, they all seemed to have this idea that they could make a difference in us," he says. "Like,

Huh! Something good could come of us. We were believed to have potential."

Carl wasn't exactly in the clear. After two stints at Mendota, he was released just before his eighteenth birthday, got married, and at age twenty was arrested for beating up a police officer. In prison, he wrote a suicide note, fashioned a makeshift noose, and was put on suicide watch in solitary confinement. While there, he began reading the Bible and fasting, and one day, he says, "something very powerful shifted." He began to believe in God. Carl acknowledges that his lifestyle falls far short of the Christian ideal. But he still attends church every week, and he credits Mendota with paving the way for his conversion. By the time he was released, in 2003, his marriage had dissolved, and he moved away from Wisconsin, eventually settling in California, where he opened his funeral home.

Carl cheerfully admits that the death business appeals to him. As a child, he says, "I had a deep fascination with knives and cutting and killing, so it's a harmless way to express some level of what you might call morbid curiosity. And I think that morbid curiosity taken to its extreme—that's the home of the serial killers, okay? So it's that same energy. But everything in moderation."

Of course, his profession also requires empathy. Carl says that he had to train himself to show empathy for his grieving clients, but that it now comes naturally. His sister agrees that he's been able to make this emotional leap. "I've seen him interact with the families, and he's phenomenal," she tells me. "He is amazing at providing empathy and providing that shoulder for them. And it does not fit with my view of him at all. I get confused. *Is that true? Does he genuinely feel for them? Is he faking the whole thing? Does he even know at this point?*"

After talking with Carl, I begin to see him as a remarkable success story. "Without [Mendota] and Jesus," he tells me, "I would have been a Manson-, Bundy-, Dahmer-, or Berkowitz-type of criminal." Sure, his fascination with the morbid is a little creepy. Yet here he is, now remarried, the father of a one-year-old son he adores, with a flourishing business. After our phone interview, I decide to meet him in person. I want to witness his redemption for myself.

*

The night before I'm scheduled to fly to Los Angeles, I receive a frantic email from Carl's wife. Carl is in police custody. He considers himself polyamorous and had invited one of his girlfriends over to their apartment. They were playing with the baby when his wife returned. She was furious and grabbed their son. Carl responded by pulling her hair, snatching the baby out of her arms, and taking her phone to prevent her from calling the police. She called from a neighbor's house instead. (Carl says he grabbed the baby to protect him.) Three misdemeanor charges—spousal battery, abandonment and neglect of a child, and intimidation of a witness—and the psychopath who made good is now in jail.

I go to Los Angeles anyway, in the naive hope that Carl will be released on bail at his hearing the next day. A few minutes before 8:30 a.m., his wife and I meet at the courthouse and begin the long wait. She is 12 years Carl's junior, a compact woman with long black hair and a weariness that ebbs only when she gazes at her son. She met Carl on OkCupid two years ago while visiting L.A. and—after a romance of just a few months—moved to California to marry him. Now she sits outside the courtroom, one eye on her son, fielding calls from clients of the funeral home and wondering whether she can make bail.

"I'm so sick of the drama," she says, as the phone rings again.

Carl is a tough man to be married to. His wife says he's funny and charming and a good listener, but he sometimes loses interest in the funeral business, leaving most of the work to her. He brings other women home for sex, even when she's there. And while he's never seriously beaten her up, he has slapped her.

"He would say sorry, but I don't know if he was upset or not," she tells me.

"So you wondered if he felt genuine remorse?"

"Honestly, I'm at a point where I don't really care anymore. I just want my son and myself to be safe."

Finally, at 3:15 p.m., Carl shuffles into the courtroom, handcuffed, wearing an orange L.A. County jumpsuit. He gives us a two-handed wave and flashes a carefree smile, which fades when he learns that he will not be released on bail today, despite pleading guilty to assault and battery. He will remain in jail for another three weeks.

Carl calls me the day after his release. "I really shouldn't have a girlfriend and a wife," he says, in what seems an uncharacteristic display of remorse. He insists that he wants to keep his family together, and says that he thinks the domestic-violence classes the court has mandated will help him. He seems sincere.

When I describe the latest twist in Carl's story to Michael Caldwell and Greg Van Rybroek, they laugh knowingly. "This counts as a good outcome for a Mendota guy," Caldwell says. "He's not going to have a fully healthy adjustment to life, but he's been able to stay mostly within the law. Even this misdemeanor—he's not committing armed robberies or shooting people."

His sister sees her brother's outcome in a similar light. "This guy got dealt a shittier hand of cards than anybody I've ever met," she tells me. "Who deserves to have started out life that way? And the fact that he's not a raving lunatic, locked up for the rest of his life, or dead is *insane.*"

I ask Carl whether it's difficult to play by the rules, to simply be *normal.* "On a scale of 1 to 10, how hard is it?" he says. "I would say an 8. Because 8's difficult, very difficult."

I've grown to like Carl: he has a lively intellect, a willingness to admit his flaws, and a desire to be good. Is he being sincere or manipulating me? Is Carl proof that psychopathy can be tamed—or proof that the traits are so deeply embedded that they can never be dislodged? I honestly don't know.

At the San Marcos Treatment Center, Samantha is wearing her new yoga pants from Target, but they bring her little joy. In a few hours, her mother will leave for the airport and fly back to Idaho. Samantha munches on a slice of pizza and suggests movies to watch on Jen's laptop. She seems sad, but less about Jen's departure than about the resumption of the center's tedious routine. Samantha snuggles with her mom while they watch *The BFG,* this eleven-year-old girl who can stab a teacher's hand with a pencil at the slightest provocation.

Watching them in the darkened room, I contemplate for the hundredth time the arbitrary nature of good and evil. If Samantha's brain is wired for callousness, if she fails to experience empathy or remorse because she lacks the neural equipment, can we say she is evil? "These kids can't help it," Adrian Raine says. "Kids don't grow up wanting to be psychopaths or serial killers. They grow up wanting to become baseball players or great football stars. It's not a choice."

Yet, Raine says, even if we don't label them evil, we must try to head off their evil acts. It's a daily struggle, planting the seeds of emotions that usually come so naturally—empathy, caring, remorse—in the rocky soil of a callous brain. Samantha has lived for more than two years at San Marcos, where the staff has tried to shape her behavior with regular therapy and a program that, like Mendota's, dispenses quick but limited punishment for bad behavior and offers prizes and privileges—candy, Pokémon cards, late nights on weekends—for good behavior.

Jen and Danny have spotted green shoots of empathy. Samantha has made a friend and recently comforted the girl after her social worker quit. They've detected traces of self-awareness and even remorse: Samantha knows that her thoughts about hurting people are wrong, and she tries to suppress them. But the cognitive training cannot always compete with the urge to strangle an annoying classmate, which she tried to do just the other day. "It builds up, and then I have to do it," Samantha explains. "I can't keep it away."

It all feels exhausting, for Samantha and for everyone in her orbit. Later, I ask Jen whether Samantha has lovable qualities that make all this worthwhile. "It can't be all nightmare, can it?" I ask. She hesitates. "Or can it?"

"It is not all nightmare," Jen responds, eventually. "She's cute, and she can be fun, and she can be enjoyable." She's great at board games, she has a wonderful imagination, and now, having been apart for two years, her siblings say they miss her. But Samantha's mood and behavior can quickly turn. "The challenge with her is that her extreme is so extreme. You're always waiting for the other shoe to drop."

Danny says they're praying for the triumph of self-interest over impulse. "Our hope is that she is able to have a cognitive understanding that 'Even though my thinking is different, my behavior needs to walk down this path so that I can enjoy the good things that I want.'" Because she was diagnosed relatively early, they hope that Samantha's young, still-developing brain can be rewired for some measure of cognitive morality. And having parents like Jen and Danny could make a difference; research suggests that warm and responsive parenting can help children become less callous as they get older.

On the flip side, the New York psychiatrist told them, the fact that her symptoms appeared so early, and so dramatically, may indicate that her callousness is so deeply ingrained that little can be done to ameliorate it.

Samantha's parents try not to second-guess their decision to adopt her. But even Samantha has wondered whether they have regrets. "She said, 'Why did you even want me?'" Jen recalls. "The real answer to that is: We didn't know the depth of her challenges. We had no idea. I don't know if this would be a different story if we were looking at this now. But what we tell her is: 'You were ours.'"

Jen and Danny are planning to bring Samantha home this summer, a prospect the family views with some trepidation. They're taking precautions, such as using alarms on Samantha's bedroom door. The older children are larger and tougher than Samantha, but the family will have to keep vigil over the five-year-old and the seven-year-old. Still, they believe she's ready, or, more accurately, that she's progressed as far as she can at San Marcos. They want to bring her home, to give it another try.

Of course, even if Samantha can slip easily back into home life at eleven, what of the future? "Do I want that child to have a driver's license?" Jen asks. To go on dates? She's smart enough for college—but will she be able to negotiate that complex society without becoming a threat? Can she have a stable romantic relationship, much less fall in love and marry? She and Danny have had to redefine success for Samantha: simply keeping her out of prison.

And yet, they love Samantha. "She's ours, and we want to raise our children together," Jen says. Samantha has been in residential treatment programs for most of the past five years, nearly half her life. They can't institutionalize her forever. She needs to learn to function in the world, sooner rather than later. "I do feel there's hope," Jen says. "The hard part is, it's never going to go away. It's high-stakes parenting. If it fails, it's going to fail big."

EVA HOLLAND

Exposure Therapy and the Fine Art of Scaring the Shit out of Yourself On Purpose

FROM *Esquire*

THE PANIC GREW with every move I made: gripping small hand-holds with suddenly sweaty palms, placing my soft rubber-soled climbing shoes onto small ledges and nubs in the granite face. My chest seized up; the fear gripping my lungs and my brain made me dizzy. I breathed loud and fast through my mouth. My brain screamed warnings at my body:

Stop! Go back!
Don't do this!
This is dangerous!
You will fall.
It will hurt.
You will get hurt!
You! Are!
Not! Safe!

It was an early May evening at the Rock Gardens, a popular climbing crag in Whitehorse, the small capital city of the Yukon Territory, where I live. By attempting to climb a steep stone wall, I was deliberately terrorizing myself, creating a situation I knew would induce something similar to a panic attack. But if I could learn to be less afraid while harnessed up and clinging to a rock face, I had decided, I might learn to control my debilitating fear of heights more generally.

That night, I managed to force my way 6 or 7 feet up a 26-foot route before I begged my climbing partner, belaying me from below, to lower me down. As my feet touched the ground, I tried to control my panting and avoided looking anyone in the eye.

Acrophobia, or extreme fear of heights, is among the most common phobias in the world: One Dutch study found that it affects as many as 1 in 20 people. Even more people suffer from a nonphobic fear of heights—they don't meet the bar to be technically diagnosed, but they share symptoms with true acrophobes like me. All told, as much as 28 percent of the general population may have some height-induced fear.

Plenty of people work around acrophobia, simply avoiding triggering situations. But seven and a half years ago, I moved to the Yukon, where many people spend their time hiking up steep mountains, climbing rock walls and frozen waterfalls, pinballing down mountain biking trails. My fear became a true liability—an obstacle between me and new friends, new hobbies, a new lifestyle. During my first full summer in Whitehorse, I panicked twice on hiking trails, curling up on the ground and refusing to move at all, or creeping along Gollum-like, on all fours, while everyone around me walked upright. It was intolerable. So last summer, I formulated a plan: I'd use the latest research to build myself a DIY cure—or, at the very least, a coping mechanism. I was going to master my fear by exposing myself to it, over and over again.

"Face your fears" is an old idea. Even its modern, clinical variation —the idea that, as a 1998 paper in the *Journal of Consulting and Clinical Psychology* put it, "emotional engagement with traumatic memory is a necessary condition for successful processing of the event and resultant recovery"—dates back more than a century, to the work of Pierre Janet and Sigmund Freud. But its codified, therapeutic application is much more recent, and it has important implications not just for people with phobias, but for those dealing with all sorts of anxiety-based conditions, from obsessive-compulsive disorder to PTSD. Facing one's fears, done correctly, could be a way forward for tens of millions of people whose anxieties control them.

I based my goals and methods of my DIY therapy program on the concept of "exposure therapy," a concept that owes its existence largely to Israeli psychologist Edna Foa, now the director of the University of Pennsylvania's Center for the Treatment and Study of Anxiety. As a postdoctoral fellow at Temple University in the early 1970s, Foa trained under Dr. Joseph Wolpe, the father of what was then known as systematic desensitization. Wolpe's work involved exposing phobic or anxious patients to the sources of their fears, mostly using "imaginal" exposure—for instance, having an arachnophobic patient imagine a spider at a distance, and then imagine the spider slightly closer, and so on—combined with relaxation techniques.

Foa's innovation was investigating whether a greater degree of "in vivo" exposure—exposure to the real fear stimulus, not just an imagined one—could improve on Wolpe's promising results. Earlier researchers had assumed such direct exposure could be dangerous for patients with phobias and anxiety disorders, but the science on that front was changing. "I started to do studies of exposure in vivo, starting not with the highest level of fear but with moderate levels, and going faster, proceeding to higher and higher situations that evoke higher and higher anxiety," Foa told me. The results, she said, were "excellent."

Exposure therapy is basically an inversion of a well-known psychological technique known as classical conditioning. If you can teach an animal to expect pain from, say, a blinking red light by repeatedly combining the light's appearance with an electrical shock until the animal reacts fearfully to the light alone, it makes sense that the twinning of stimulus and fear can be *unraveled* too. Show the animal the red light enough times without an accompanying shock, and eventually it will no longer fear the light—a process known as extinction. I was determined to extinguish my fear by proving to myself that I could climb a cliff.

If I was afraid of heights as a small child, I don't remember it. I never climbed trees, and I was uncomfortable when my friends and I clambered up to sit on top of the monkey bars on the playground. But I was a timid kid in general—I once told my mom that I never ran as fast as I could in school races, for fear of losing control and falling—so all that was of a piece with my personality at that time.

In my first clear memory of feeling afraid of heights—not just afraid, but *terrified*—I am fifteen years old. It was the summer after ninth grade, and I'd signed up to spend a week sailing on an old-fashioned ship on Lake Ontario with a dozen other teens. I loved everything about life on board that ship: sleeping in my narrow metal bunk below deck; waking in the middle of the night to stand watch, peering out at the endless darkness; lounging on sunny afternoons in the net that hung below the carved bow. On deck, we wore harnesses around our chests, fitted with a short rope ending in a heavy metal clip. In very rough weather, or if we were climbing the mast to adjust the sails, we were meant to clip ourselves in, just in case.

The problem came the first time I tried to climb the mast—to "go aloft," in sailing terminology. I got partway up, moving my clip as I went, fighting panic with each step on the ladder-like holds. Then I froze. I couldn't stop staring at the wooden deck swaying below me, couldn't stop picturing my body splattering against it, my bones shattering, my blood running into the lake.

The ship's "officers"—our camp counselors—managed to coax me down, and I never went aloft again. Everyone was kind to me about my failure, but there was no point in coming back the following year. A sailor who can't adjust the sails in a pinch isn't much use.

After that, my fear went dormant again for nearly a decade. It resurfaced after grad school, while I was backpacking with friends in Europe. I'd developed a fascination with the art and architecture of old churches, and we hit cathedral after cathedral across the southern half of the continent. We visited a few cupolas, and I gritted my teeth going up and down the narrow stone stairways. But I didn't truly panic until Florence.

I'd made it to the top of the legendary Duomo and was breathing deeply, trying to stay calm and enjoy myself as I looked out over the city's terracotta rooftops. The famous steep red dome of the cathedral curved away below me, and as I glanced down at it, suddenly all I could think about was how it would feel to tumble over the flimsy metal railing in front of me, to slide down over those red tiles toward the drop-off. I couldn't breathe.

The viewing platform was crowded with tourists. I pushed through them to the wall and slid down with my back against it, put my head between my knees to block out the view, and hyperventilated through my tears. My friends found me there, eventually talked me to my feet, and held my hands while we inched back down the twisting staircase to safety and solid ground. We didn't visit any more cathedral towers after that.

In the years since that humiliating incident, I've tried to figure out why I react to heights—specifically exposed heights; I'm generally fine in enclosed spaces, like elevators and airplanes—the way I do. Phobias can often derive from traumatic experiences, or even observations of others' traumatic experiences, early in life. But it turns out that the acrophobia is different. If I'm anything like the subjects of recent research, I have measurably sub-par control over my body's movement through space, as well as an over-dependence on visual cues—which are distorted by heights—to manage my movement through the world. In other words, I am afraid of falling from heights because I am more likely than other people to fall from heights.

For a 2014 paper in the *Journal of Vestibular Research,* a team of German scientists studied the eye and head movements of people who are afraid of heights, plus a control group, as they looked over a balcony. They found that their fearful subjects tended to restrict their gazes, locking their heads in place and fixing their eyes on the horizon rather than looking down or around at their surroundings. That description will ring true to anyone who's ever felt afraid of heights, or tried to counsel someone who is: *Don't look down. Whatever you do, don't look down.*

So, ironically, I fix my gaze to the horizon as a defense mechanism against my fear, but because that fear is rooted in my over-reliance on visual cues, restricting my range of vision can only make things worse. It's a cycle: My brain knows that my body is bad at navigating heights, so it sends out fear signals as a warning. My body shuts down in response, which only increases the likelihood that I will actually harm my klutzy self. And thus a once-rational response to a reasonable concern feeds on itself, growing and spreading to the point where I can hardly stand on a sturdy stepladder.

A few weeks after that first outing in May, I was back at the Rock Gardens. I'd been making sporadic attempts to face my fears for years, but now I intended to be more systematic about my efforts, and to document them as I went.

The route I was attempting was a beginner's climb, laughably easy for most people with any experience. And it came with a cheat option: a detour of a few feet to the right, into a wide crack between two rock faces, made it even simpler. But to get to the crack and the easiest way up, I had to make one slightly tricky move. I would have to step forward with my left foot, balance the toe of my shoe on a small nub, shift all my weight briefly to that left toe, then swing my right foot over and across to the next proper ledge—all without any handholds for balance.

My climbing partner stood below me, holding the other end of the rope that secured me to the bolted metal anchors at the top of the climb. If I fell, she would pull down on the rope, stopping me before I'd plummeted more than a foot or two. Climbing on top rope, as it's known, involves almost no real risk. But my lungs constricted anyway, and I fought to squelch my dizziness and panic. From the ground, my friends encouraged me: *Trust your shoes, trust your feet. This will be fine. You can do this.*

Finally, I took a deep breath, stepped forward, shifted my weight from one foot to the next, and made it across. I fumbled above my head for handholds to steady myself, then grinned and tried to breathe. For a moment while I was in motion, I had felt weightless, in control. Unafraid. Now the fear came seeping back as I continued climbing, scrambling through the loose dirt that had collected on the ledges and lumps of rock in the crack. I finished the climb, but raggedly, fending off panic the whole way. It was a good start, but as my belayer lowered me back down to the ground, I knew I had a long way to go.

We don't know exactly what happens in the brain during the extinction process. As Foa puts it, "is it that you erase the connections" between stimulus and fear, "or that you replace them with a new structure?" Her hypothesis is that exposure therapy trains the brain to create a second, competing structure alongside the traumatic one. The new structure, she explained, "does not have the fear, and does not have the perception that the world is entirely dangerous and that oneself is entirely incompetent."

That was why my panicked success in the Rock Gardens that day was really no success at all. I had climbed the wall, sure, but I had failed to convince my brain to build a new structure. Repeatedly terrorizing myself wouldn't solve my fear; it wasn't enough to scramble through with wild eyes and a pounding heart. I had to learn to stay calm.

Perhaps the most transformative application of exposure therapy is using it not to combat specific phobias, or even broader anxiety-based disorders, but post-traumatic stress disorder. In 1980, PTSD was included for the first time in the *Diagnostic and Statistical Manual of Mental Disorders*. In the decades since, our understanding of the disorder has grown, and so has our grasp of its staggering reach. We now know that PTSD affects not just soldiers and civilians emerging from war but also drone operators who've never left their home base; first responders from beat cops to search-and-rescue volunteers operating out of luxurious mountain resorts; survivors of car wrecks, assaults, and less obvious forms of trauma.

But back in the early 1980s, "we didn't have any studies on PTSD," Foa said. "And I thought, well, this is an anxiety disorder, there is no reason why we cannot adapt the treatment, the exposure therapy treatment, to PTSD." You can't re-expose someone to a rape or a bomb, so Foa settled on a program of imaginal exposure for the traumatic memory itself but in vivo exposure to the secondary effects: the patient's avoidance behaviors, which can perpetuate trauma's power. In sessions with therapists, patients would confront the memory using imaginal exposure. Their "in vivo" exposure came as homework: going to places that reminded them of the trauma, or to safe places they perceived as dangerous. Sometimes that meant walking a downtown street at night after a violent assault or going to malls again after a mass shooting.

Throughout the 1990s, Foa's team taught other groups of therapists how to administer what she called prolonged exposure therapy (or PE), and how to monitor the results. They found that PE was effective in almost 80 percent of patients: between 40 and 50 percent became essentially symptom-free, while 20 to 30 percent still had some recurring symptoms but were much improved. "We're not 100 percent successful," she said, "but no treatment is." She launched PE into the wider world with a series of papers in the late 1990s, and within a few years the program had become the gold standard for treatment of anxiety disorders and PTSD. In 2010, Foa was named one of *Time*'s 100 most influential people.

"No one is doing more" to end the suffering caused by PTSD, the magazine declared.

An estimated 8 million American adults experience PTSD every year. Nineteen million more deal with specific phobias, 6 million with panic disorders, 7 million with generalized anxiety disorder, and more than 2 million with OCD. The Anxiety and Depression Association of America estimates that only one-third of anxiety-disordered patients receive treatment. Now, researchers are exploring whether pharmaceuticals can enhance the effectiveness of exposure therapy, while others have applied variations of PE to grief, depression, eating disorders, and beyond.

Compared to living with PTSD or broader anxiety disorders, my fear of heights is trivial. It doesn't keep me awake at night, or ruin my relationships, or bleed into every area of my life. If I moved back to the flatlands and avoided high-rise balconies, dodging my symptoms by practicing avoidance, I would hardly notice it.

Still, it can limit me. I would have liked to climb that mast high into the rigging, to enjoy the view over Florence. Sometimes I get scared on bridges or balconies, and I have still never climbed a tree. Taken individually those are all tiny things, but they add up to a feeling of helplessness: my choices are not entirely my own.

The rock was cold enough to numb my fingers. It was October 2, and I was on my eighth and final climbing excursion of the season, before winter set in. All summer, I had gone climbing every time someone with the necessary expertise and gear was willing to take me along. I had tried to systematize my outings, repeating the same routes to see if I could get farther, and stay calmer, each time.

In previous years, I would have pushed myself until my panic was unbearable, hoping that I could pop it like a soap bubble if only I tried hard enough. But now my strategy was to go only as far up as I could without paralysis setting in. The goal was to build up the alternate structure in my brain that said, "This is okay. You are safe," then come down before the old structure could assert itself, and hope to get a foot or two farther next time around.

For this last outing, three friends and I were at Copper Cliffs, a crag in Whitehorse's semi-industrial backyard: once a booming copper-mining area, now a maze of quarries and mountain biking trails and small, shallow lakes. I was climbing Anna Banana, a short, beginner-friendly, 16-foot route up one side of an arête, a sharp wedge of rock protruding from the main cliff face. My first steps had been on easy footholds, gaps cutting into the leading point of the wedge, and I had no trouble until my feet were seven and a half, eight feet off the ground. I stalled out there, my right foot resting on a good ledge just around the corner of the arête while my left toe was tucked into a little cubbyhole a foot below. To continue, I had to pull my left leg up several feet, to the next good hold.

I raised my arms and patted the rock above my head, blindly seeking out handholds that I could use to pull myself up higher, to give my left foot a fighting chance. I tend to trust my hands and arms first, even though my legs are exponentially stronger: we're less accustomed to trusting a narrow toehold than a fist clamped around something solid. But I didn't find what I was looking for, so instead I spread my arms out wide and locked my fingers around the best stabilizing holds I could reach. Then I pushed off with all my weight on my right foot, pulled my arms tight to keep me close to the rock face, and scraped my left foot up the wall until I found the next hold, just as my right toe lost contact with the rock. I balanced there for a moment, then raised my hands to holds suddenly within my reach and pulled up my dangling right foot.

I had done it. More importantly, I had done it calmly and coolly, without needing extra minutes to fight off panic, without groaning and moaning before I gave it a try. My belayer lowered me down so I could climb up and do it again—more confidently, with even less hesitation. This time I kept going, through a series of easy moves to the top of the route, where I reached up and smacked the anchor bolts in triumph: a touchdown spike. I did a quick mental survey of my body: my breathing was steady, my head clear. For today, at least, I had successfully redirected my brain to reject fear.

Months later, I'm still working on training my brain. I've kept climbing through the winter, at big indoor gyms in San Francisco and Vancouver and on small, homemade climbing walls here at home; in local schools and in a friend's basement. By my standards, I've made substantial progress. These days my chest doesn't constrict and my pulse doesn't start to pound in my ears until I'm much higher off the ground: 6, 8, 10 feet. Sometimes I can complete an entire short route without feeling afraid at all.

I've started applying the basic ideas behind exposure therapy in other areas of my life too. So often, whether in our careers or our athletic endeavors or even our love lives, we're encouraged to "take the plunge," to "push our limits," to "go big or go home." But my DIY climbing therapy has taught me the value of care, of caution, of building up your abilities and endurance slowly to reach a larger goal. Taking the plunge has its place, but sometimes it's enough to immerse yourself toe by toe.

KATHRYN SCHULZ

Fantastic Beasts and How to Rank Them

FROM *The New Yorker*

CONSIDER THE YETI. Reputed to live in the mountainous regions of Tibet, Bhutan, and Nepal. Also known by the alias Abominable Snowman. Overgrown, in both senses: 8 or 10 or 12 feet tall; shaggy. Shy. Possibly a remnant of an otherwise extinct species. More possibly an elaborate hoax, or an inextinguishable hope. Closely related to the Australian Yowie, the Canadian Nuk-luk, the Missouri Momo, the Louisiana Swamp Ape, and Bigfoot. Okay, then: on a scale not of zero to ten but of, say, leprechaun to zombie, how likely do you think it is that the yeti exists?

One of the strangest things about the human mind is that it can reason about unreasonable things. It is possible, for example, to calculate the speed at which the sleigh would have to travel for Santa Claus to deliver all those gifts on Christmas Eve. It is possible to assess the ratio of a dragon's wings to its body to determine if it could fly. And it is possible to decide that a yeti is more likely to exist than a leprechaun, even if you think that the likelihood of either of them existing is precisely zero.

In fact, it is not only possible; it is fun. Take the following list of supernatural beings:

- Angels   - Giants

- Demons   - Pegasus

- Dragons   - Centaurs

– Pixies	– Unicorns
– Ghosts	– Tooth fairy
– Harpies	– Phoenix
– Elves	– Werewolves
– Mermaids	– Vampires
– Loch Ness monster	– Genies
– Leviathan	– Zombies

Never mind, for now, whether or not you actually believe in any of these creatures. We are interested here not in whether they are real but in to what extent they seem as if they could be. Your job, accordingly, is to rank them in order of plausibility, from most likely (No. 1) to least likely (No. 20). Better still, if you are in the mood for a party game this Halloween season, try having a lot of people rank them collectively. I guarantee that this will produce a surprising amount of concord—who among us could rank the tooth fairy above the Leviathan?—as well as a huge amount of impassioned disagreement. The Loch Ness monster will turn out to have a Johnnie Cochran–level defense attorney. Good friends of yours will say withering things about mermaids.

What's odd about this exercise is that everyone knows that "impossible" is an absolute condition. "Possible versus impossible" is not like "tall versus short." Tall and short exist on a gradient, and when we adjudge the Empire State Building taller than LeBron James and LeBron James taller than Meryl Streep, we are reflecting facts about the world we live in. But possibility and impossibility are binary, and when we adjudge the yeti more probable than the leprechaun we aren't reflecting facts about the world we live in; we aren't reflecting the world we live in at all. So how, exactly, are we drawing these distinctions? And what does it say about our own wildly implausible, unmistakably real selves that we are able to do so?

In the fourth century B.C., several hundred years after the advent of harpies and some two millennia before the emergence of dementors, Aristotle sat down to do some thinking about supernatural occurrences in literature. On the whole, he was not a fan; in his *Poetics,* he mostly discouraged would-be fabulists from messing around with them. But he did allow that, if forced to choose, writers "should prefer a probable impossibility to an unconvincing possibility." Better for Odysseus to return safely to Ithaca with the aid of ghosts, gods, sea nymphs, and a leather bag containing the wind than for his wife, Penelope, to get bored with waiting for him, grow interested in metalworking, and abandon domestic life for a career as a blacksmith.

As that suggests, for a possible thing to seem plausible it must be reasonably consistent with our prior experience. But what makes an impossible thing seem plausible? In a convoluted passage in the *Poetics,* Aristotle tells us that if an impossible thing would "necessarily" require something else to occur along with it, you should put that second thing in your story too, because then your readers will be more likely to believe the first one. In other words, even something that is factually impossible can be logically possible, and how closely that logic is followed will affect how plausible a supernatural being seems.

There's a reason Aristotle addressed this advice to writers and artists. Unlike most of us, they have practical motives for wondering how best to make imaginary things seem convincing, a problem that must be solved as much for *Vanity Fair* as for *A Wrinkle in Time.* Accordingly, creative types have done an unusual amount of thinking about plausible impossibility. In the 1790s, for instance, Samuel Taylor Coleridge set out to write a series of poems about "persons and characters supernatural." To do so, he knew, he had to make the fantastical seem credible — "to procure for these shadows of imagination," he wrote, in a soon to be famous phrase, a "willing suspension of disbelief."

Coleridge was excellent at inducing a suspension of disbelief. That's why we are as gripped by "The Rime of the Ancient Mariner" as the wedding guest within the poem who can't tear himself away from the sailor's tale — even though the tale itself is an outrageous one involving a magical albatross, a terrible curse, and a ship crewed by ghosts. Yet Coleridge was vague about explaining how exactly he did it. His only advice for making impossible things seem believable was to give them "a semblance of truth."

A little more than a hundred years later, a very different kind of artist got somewhat more specific. Although Walt Disney is best remembered today for his Magic Kingdom, his chief contribution to the art of animation was not his extraordinary imagination but his extraordinary realism. "We cannot do the fantastic things, based on the real, unless we first know the real," he once wrote, by way of explaining why, in 1929, he began driving his animators to a studio in downtown Los Angeles for night classes in life drawing. In short order, the cartoons emerging from his workshop started exhibiting a quality that we have since come to take for granted but was revolutionary at the time: all those talking mice, singing lions, dancing puppets, and marching brooms began obeying the laws of physics.

It was Disney, for instance, who introduced to the cartoon universe one of the fundamental elements of the real one: gravity. Even those of his characters who could fly could fall, and, when they did, their knees, jowls, hair, and clothes responded as our human ones do when we thump to the ground. Other laws of nature applied too. Witches on broomsticks got buffeted by the wind. Goofy, attached by his feet to the top of a roller-coaster track and by his neck to the cars, didn't just get longer as the ride started plunging downhill; he also got skinnier, which is to say that his volume remained constant. To Disney, these concessions to reality were crucial to achieving what he called, in an echo of Aristotle, the "plausible impossible." Any story based on "the fantastic, the unreal, the imaginative," he understood, needed "a foundation of fact."

Taken together, Disney's foundation of fact and Coleridge's semblance of truth suggest a good starting place for any Unified Theory of the Plausibility of Supernatural Beings: the more closely such creatures hew to the real world, the more likely we are to deem them believable. But the real world is enormous, wildly heterogeneous, extraordinarily complicated, and, itself, often surpassingly strange. So if, indeed, the most plausible supernatural creatures are those which most resemble reality, the question becomes: which part?

The obvious candidate, at first glance, is the animal kingdom. Supernatural creatures are, after all, creatures, and we infer from them, or impose upon them, all kinds of biological characteristics. Like their natural counterparts, they can be organized by taxon (cervid, like the white stag; caprid, like the faun; bovine, like the Minotaur; feline, like the sphinx), or by habitat (alpine, like yetis;

woodland, like satyrs; cave-dwelling, like dragons; aquatic, like mermaids). Given this tendency to situate unnatural beings in the natural world, it seems conceivable that our judgments about their plausibility might reflect how well they conform to the constraints of modern biology.

If that's the case, our friend the yeti should rank very high on the believability scale. So, too, should giants, elves, unicorns, ogres, imps, sea monsters, and pixies. By the same token, this biological theory would deal a credibility blow to angels, demons, fairies, vampires, and werewolves, plus all those creatures assembled, as by an insane taxidermist, from the separate parts of real species: mermaids, griffins, centaurs, chimeras, sphinxes. It would also undermine the plausibility of fire-breathing dragons, there being no analogue in nature to a Zippo. In fact, biological limitations cast doubt on dragons in another way as well, since four legs plus two wings is not a naturally occurring configuration—a bummer also for harpies, griffins, gargoyles, and Pegasus.

If you couldn't make it through that paragraph without starting to formulate an objection, you already know the first problem with this theory: it invites a lot of quibbling over what is and isn't biologically feasible. As defenders of the supernatural will be quick to point out, many arthropods have six limbs; squids, skunks, bombardier beetles, and plenty of other real creatures spew strange things; nature sometimes contrives to recombine old animals in new ways (see the half-striped zedonk—part zebra, part donkey —or the recent emergence of the coywolf: part coyote, part wolf); and, considering the many kinds of metamorphoses exhibited by animals—tadpole to frog, caterpillar to butterfly, baby-faced to bearded—how far-fetched is it, really, for a bat to turn into a man?

Indeed, some fantastical creatures seem positively ordinary compared with the more byzantine products of 4 billion years of evolution. Consider the giant oarfish, a 36-foot-long behemoth with a silver body, a bright-red mane, and a tendency to hang out in the ocean vertically, like a shiny piscine telephone pole. Or consider the blue glaucus, an inch-long hermaphroditic sea slug capable of killing a Portuguese man-of-war—a beast 300 times its size —and then storing its poison for later use, including on humans.

Given so much natural extravagance, it's not surprising that the real and the unreal are sometimes mistaken for each other. In 1735, when Carl Linnaeus organized all the species in the world

into one vast taxonomy, he included a section on "Animalia Para-
doxa": creatures, common in folklore and myth or attested to by
far-flung explorers, that he felt compelled to itemize yet deemed
unlikely to exist. Among these were the manticore (head of a man,
body of a lion, spiky tail), the lamia (head of a man, breasts of a
woman, body of a scaly cow), and the Scythian lamb (like a regular lamb, except it grows out of a stalk in the ground)—but also,
arrestingly, the antelope and the pelican. Conversely, a contributor to *This American Life* once recounted the experience of asking
a group of strangers at a party, in all sincerity, whether unicorns
were endangered or extinct. One sympathizes. Consider the gi-
raffe. Consider the kangaroo.

On top of all this, the biological theory of plausibility also suf-
fers from a graver problem: its predictive powers are faulty. By its
logic, many creatures that we find highly believable should instead
rank near the bottom of the list. Angels, for instance, are physi-
ologically unlikely: in addition to being able to fly (fine for birds,
unheard of in hominids), they manifest a particularly extreme version of the limb problem, since, per various sources, they have
not just two but in some cases hundreds of wings. Demons present the same basic difficulties, as do fairies, and ghosts defy pretty
much *every* biological principle: among other problems, they have
no substance, require no sustenance, and do not decay or die. Yet
given that seven out of ten Americans believe in angels, six out of
ten believe in demons, and almost half believe in ghosts, it seems
safe to assume that, on the scale of plausibility, such creatures out-
rank giants and unicorns.

So much for biology as the basis of our unified theory. But we
can resolve at least some of these problems by modifying our hypothesis slightly. Perhaps we don't care how much supernatural
creatures resemble the animal kingdom in general; perhaps we
only care how much they resemble *us*. This mirror theory of plausibility would still account for the high ranking of yetis, which,
aside from not existing, are not so different from *Homo sapiens*.
(Back in 2004, when scientists discovered an extinct species of an
unusually small hominid on an island in Indonesia, a senior editor at *Nature* took the occasion to speculate that stories about yetis might reflect an extinct Himalayan species on the other end
of the size spectrum.) The mirror theory would also explain the
perceived plausibility of angels and demons, which, as presented

in myth and literature, resemble exaggerated humans in our best and worst incarnations: moral giants and moral elves. And it would explain why vampires and werewolves, which should rank low on the list, what with the impossibility of radical metamorphosis, generally rank quite high. When they are not busy sprouting wings and fur, after all, such creatures look nearly indistinguishable from us.

On the other hand, this theory leads us quickly into ontological problems: are we humans more like mermaids, or more like ghosts? Worse, like the biological theory of plausibility, it fails to account for some of our intuitions about supernatural beings. Why, for example, would a centaur, which is 50 percent human, strike us as less plausible than a unicorn, which is 0 percent human? And what are we to make of natural-born humans who are able to do supernatural things, à la Shakespeare's Prospero, Hermione Granger, or that menace of Camelot Morgan le Fay?

This last category of being opens up a whole new can of worms. Magical creatures exist in a universe of magical powers, which themselves range wildly in probability and are not evenly distributed among the population. To understand our intuitions about plausibility, then, we need to look beyond entities to actions. For supernatural creatures, as for the rest of us, it might be that what matters most is not what we are but what we do.

What *do* supernatural creatures do? In many cases, not much. Somewhat strangely, not every magical being has magical powers. Some, like Santa Claus and the tooth fairy, mostly just have chores. Others merely hang around looking unusual; the yeti and Nessie just lurk; the Leviathan lurks, too, largerly; the record is mixed on giants, which in some accounts live on clouds but in most are just enormous and crabby. Wraiths only scare people, centaurs only awe people, and unicorns, aside from some healing properties in their horns, akin to the antibiotics in frog skin, only attract virgins —which, power-wise, puts them at the same level as boy bands. For these and many other supernatural creatures, their supernaturalness inheres chiefly in the fact (or the non-fact) of their existence.

Others, however, can do flatly impossible things. Fairies, by most accounts, can turn invisible, tell the future, and shape-shift. Ghosts can shrink, expand, time-travel, and walk through walls. Vampires can command the dead, summon storms, control lesser animals like bats and wolves, and—barring certain interventions with stakes or sunlight—live forever. Various other entities can, through their own powers or via potions, amulets, and spells, likewise achieve the unachievable: levitate, teleport, transmogrify, read minds, talk to animals, and, by occult means, charm, confuse, possess, haunt, hex, heal, or kill.

Like supernatural creatures, such powers can be ranked in terms of plausibility. Which seems more likely to work: Harry Potter's apparating ability or Obi-Wan Kenobi's Jedi mind trick? If you ask me, it's obviously the mind trick, with its real-life analogies of charisma and hypnosis, not to mention its failure to defy any major laws of physics. On the other hand, apparating—vanishing from one place and appearing in another—strikes me as more plausible than time travel, possibly because we have many ways to move through space but only one way to move through time.

You can play this game forever, with any given set of magical powers. Controlling the elements, for instance, seems considerably harder than controlling an animal (unless, perhaps, it is a cat)—but, if you *are* going to try to control the elements, summoning a breeze seems easier than turning night to day. If you're going to work magic on your own body, becoming invisible seems more plausible than transmogrifying, perhaps because of the abundance of everyday ways to conceal ourselves. Yet, if transmogrification is going to occur, I'd wager that it is easier to turn oneself into a wolf than one's enemy into a toad.

As it happens, intuitions like these are broadly shared—a fact we know because, speaking of implausible things, two cognitive scientists at the Massachusetts Institute of Technology have shown it. Normally, Tomer Ullman studies our commonsense beliefs about physics and psychology, while his colleague John McCoy studies judgment and decision-making. Together, however, they figured that looking at how we reason about supernatural powers might shed light on how we reason about the real world. To that end, in 2015 they asked 200 people, ranging in age from eighteen to eighty-three, to rank 10 magic spells in order of difficulty. Since amphibians in magic have roughly the same status as rodents in science, all the spells featured things a sorcerer could do to a frog: conjure it into existence, conjure it out of existence, teleport it, levitate it, change its color, double its size, turn it into two frogs, turn it into a mouse, turn it to stone, and turn it invisible.

The results help explain why I am dubious about apparating. Overall, the subjects felt that spells were more difficult when they violated "more fundamental principles of intuitive physics." What makes a principle of physics fundamental, in this case, is how early in our cognitive development we acquire it. For instance, we are born with an understanding of object permanence, and the two spells that violated it, by conjuring a frog into or out of existence, were ranked the most difficult. Similarly, we learn in infancy that objects have what developmental psychologists call "kind-identity" —they stay themselves—which may explain why the next-hardest spell involved turning a frog into a mouse. The two easiest spells, by comparison, entailed changing a frog's color and levitating it, results that reflect our awareness that both color and location are transient rather than fixed features of the physical world.

To further plumb our intuitions about supernatural powers, McCoy and Ullman ran a second study, which asked the same questions but changed one of two things: either the target of the spell (Is it harder to conjure a frog or a cow?) or the extent of its power (Is it harder to levitate a frog 1 foot or 100 feet?). Resoundingly: a cow; 100 feet. These findings are striking, since levitating something 99 extra feet does not violate any additional principles of physics. Nor does conjuring a cow instead of a frog. So why would those variants seem more challenging?

Happily, two other cognitive scientists, Andrew Shtulman and Caitlin Morgan, of Occidental College, have addressed that question. (Full disclosure: my sister, a cognitive scientist at MIT, was Shtulman's postdoctoral adviser and has worked with McCoy and Ullman.) Last year, Shtulman and Morgan gave people pairs of magic spells and asked them to determine which one in each pair was more difficult. In every pair, both spells violated the same fundamental principle of physics, biology, or psychology, but each varied in how much it violated a secondary one. For instance, physics dictates that you can't walk through anything solid, no matter what it's made of, but also that materials differ with respect to properties like density and hardness. So which seems more difficult: walking through a wall made of stone or a wall made of wood?

Overwhelmingly, the subjects chose stone. They also determined that it would be harder to levitate a bowling ball than a basketball, and harder to grow an eye than a toe. Since levitation is categorically impossible, it shouldn't matter that heavier objects, like bowling balls and cows, are harder to lift. But, as Aristotle understood, it does. According to Shtulman and Morgan, that's because our understanding of causation—our sense of which things make other things happen—is not a series of separate if-then statements but a vast interconnected web, which continues to govern our intuitions even when one particular strand snaps. "Severing one link in a causal network," they write, "still leaves the rest of the network intact." And the more links you sever, the more powerful—or, put differently, the less probable—your magic seems.

Perhaps, then, the solution we seek is mathematical: tally up all the fundamental principles violated by a supernatural creature and its powers and—voilà, we'll know where it stands in the hierarchy of likelihood. Call this the parsimony theory of plausibility: the fewer laws something violates, the more credible it will seem. The yeti, for instance, doesn't really violate any natural laws at all. Vampires, by contrast, violate everything from the fact that things of substance cast shadows to Meteorology 101.

This parsimony theory is simple, elegant, and, unfortunately, wrong. If it were correct, we'd all find gnomes, whose only distinguishing characteristics are diminutiveness, avarice, and a preference for living underground, considerably more plausible than ghosts. Yet ghosts, despite their utter disregard for biology and physics, persist in seeming highly believable. Part of that might be explained by our existential condition: most of us feel that we have a core self, separate and separable from our body, and most of us find it hard to accept that we will someday cease to exist.

Part of it, however, might be explained by one final theory of supernatural plausibility. Consider a defense that my sister once mounted on behalf of the likelihood of fairies. Small impossible things, she contended, are more believable than large impossible things, because they could more easily exist without us noticing them. That argument isn't based on our beliefs about physics or biology; it's based on epistemology. From infancy on, we are extraordinarily sensitive to patterns of evidence (in fact, that's how we acquire many of our beliefs about physics and biology), so it seems reasonable to think that evidence also determines our judgments about fantastical beings.

Of course, it also seems *unreasonable* to think that, since it's unclear how we would find evidence for the existence of nonexistent creatures. In its absence, we can make do, as my sister did, with a good reason for why we haven't found it, a strategy that lends plausibility not only to fairies in their tininess but also to ghosts and other creatures capable of vanishing. (It also gives us a reason, finally, to object to the yeti: if it existed, we should have found proof by now.)

Alternatively, we can accept attestation as a form of evidence—which, across domains, we do all the time, since many of our convictions about the world concern things we ourselves will never observe. Our sensitivity to attestation explains why culture has such a potent influence on our intuitions about the supernatural, which wouldn't be the case if those intuitions were governed chiefly by biology or physics. It is why one community is more likely to believe in fairies and another in zombies, and why, with churches peddling a more palliative version of Christianity, demons have declined in plausibility vis-à-vis angels. And it is why, if you're European American, you're more likely to believe in a vampire than in the coffin-dwelling, night-roaming, life-force-sucking Chinese *jiangshi,* even though, on the basis of their characteristics, there is not, so to speak, a lot of daylight between them.

Patterns of evidence, a grasp of biology, theories of physics: as it turns out, we need all of these to account for our intuitions about supernatural beings, just as we need all of them to explain any other complex cultural phenomenon, from a tennis match to a bar fight to a bluegrass band. That might seem like a lot of intellectual firepower for parsing the distinctions between fairies and mermaids, but the ability to think about nonexistent things isn't just handy for playing parlor games on Halloween. It is utterly fundamental to who we are. Studying that ability helps us learn about ourselves; exercising it helps us learn about the world. A three-year-old talking about an imaginary friend can illuminate the workings of the human mind. A thirty-year-old conducting a thought experiment about twins, one of whom is launched into space at birth and one of whom remains behind, can illuminate the workings of the universe. As for those of us who are no longer toddlers and will never be Einstein: we use our ability to think about things that aren't real all the time, in ways both everyday and momentous. It is what we are doing when we watch movies, write novels, weigh two different job offers, consider whether to have children.

As that last example suggests, perhaps the most extraordinary thing about this ability is that we can use it to nudge the impossible into the realm of the real. We stare at the sky, watch a seagull bob on a thermal, build wax wings and then fixed wings and then Apollo XI. We dream of black presidents and female scientists; we dream, still, of self-driving cars, a cure for cancer, peace in the Middle East. These last things are interestingly like dragons and also interestingly unlike dragons, in ways that suggest that we may be wise, after all, to treat impossibility as something other than an absolute condition. Alone among all the creatures in the world, we can think about fantastical things and, at least some of the time, bring them into being.

Yet, in the end, what's most remarkable is not that our fantasies contain so much reality; it is that our reality contains so much fantasy. Most of us understand that our perceptual systems, far from passively reflecting the world around us, actively sort, select, distort, ignore, and alter a huge amount of information in order to construct reality as we experience it. But reality as we experience it also departs from actual reality in deeper ways. In actual reality, space and time are inseparable, and neither one behaves anything like the way we perceive it; nor does light, and nor does gravity, and, in all likelihood, nor does consciousness. Yet all the while we go on experiencing space like a map we can walk on, time like a conveyor belt we travel on, ourselves as brimming with agency, our lives as mattering urgently.

That world, the one we inhabit every day of our lives, is a yeti —a fantastical thing constructed out of bits and pieces of reality plus the magic wand of the mind. If we could hand it over to some superior being for consideration, it might not even rank very high on the scale of plausibility. Then again, plausibility itself might not rank very high on the scale of qualities we prize. Better, perhaps, to know that what we feel in our happiest moments has some truth to it: life is magical.

Contributors' Notes

Other Notable Science and Nature
Writing of 2017

Contributors' Notes

Ross Andersen is a senior editor at *The Atlantic*.

Rebecca Boyle is an award-winning freelance journalist who focuses on how things work and the ways humans try to understand the universe. Her magazine writing covers astronomy and physics, geoscience and climate change, and the history of science. She is a contributing writer for *The Atlantic*, and her work regularly appears in *FiveThirtyEight, New Scientist, Quanta,* and many other publications for adults and kids. Rebecca grew up in Denver and dreams of the pristine night skies over the Rocky Mountains. She lives in St. Louis with her family.

Sophie Brickman is a writer and editor based in New York City. Her work has been published in *The New Yorker,* the *New York Times,* and the *San Francisco Chronicle,* among other places.

Kenneth Brower is the son of the pioneering environmentalist David Brower. His first memories are of the Sierra Nevada and the wild country of the American West. He is a freelance writer and the author of many books and magazine articles on the environment and natural history. His work has taken him to all the continents. He lives in Berkeley, California.

Jacqueline Detwiler is a senior editor at *Popular Mechanics.* She has also written for *Wired,* Esquire.com, *Entrepreneur,* and other publications. She holds an M.A. in psychology and neuroscience from Duke University and is the host of *Popular Mechanics'* podcast, *The Most Useful Podcast Ever.*

Ceridwen Dovey regularly contributes nonfiction to newyorker.com, and is the author of the novels *Blood Kin* and *In the Garden of the Fugitives* and the short-story collection *Only the Animals.* She lives in Sydney, Australia.

Susannah Felts is a writer, editor, and cofounder/co-director of The Porch, a literary arts organization in Nashville, Tennessee. She is the author of a novel, *This Will Go Down on Your Permanent Record,* and her essays and fiction have appeared in publications such as *Guernica, Longreads, Catapult, Oxford American,* the *Sun, Hobart, Smokelong Quarterly,* and others. She's at work on a novel.

Douglas Fox (www.douglasfox.org) is a freelance journalist who writes extensively on Earth, the Antarctic, and polar sciences. His stories have appeared in *Scientific American, National Geographic, Esquire, Virginia Quarterly Review, High Country News, Discover, Nature, Slate,* the *Christian Science Monitor,* and other publications. Stories by Doug have garnered awards from the American Society of Journalists and Authors (2011), the National Association of Science Writers (2013), the American Geophysical Union (2015), the Society of Environmental Journalists (2016), and the American Association for the Advancement of Science (2009, 2017). Doug is a contributing author to *The Science Writers' Handbook.*

Barbara Bradley Hagerty is the *New York Times* bestselling author of *Life Reimagined: The Science, Art, and Opportunity of Midlife* and *Fingerprints of God: The Search for the Science of Spirituality.* She regularly contributes to *The Atlantic* and NPR. Barb worked for NPR for 19 years (1995–2014), covering law and religion. She has received the American Women in Radio and Television Award (twice), the National Headliners Award, and the Religion Newswriters Association award for radio reporting. She lives in Washington, D.C., with her husband, Devin.

Eva Holland is a freelance writer based in Canada's Yukon Territory. She's a correspondent for *Outside* magazine, and her work has also appeared in *Wired, Pacific Standard, Hakai,* and many other outlets in print and online. She is working on a book about the science of fear.

Steven Johnson is the bestselling author of 11 books on science, technology, and the history of innovation, including *The Ghost Map, Where Good Ideas Come From,* and *Wonderland.* He is also the host and co-creator of the PBS series *How We Got to Now.* He writes regularly for *The New York Times Magazine* and *Wired.* He lives in Brooklyn, New York, and Marin County, California, with his wife and three sons.

Caitlin Kuehn is from small-town Wisconsin, where she graduated with a B.S. in biology and, for one whole exciting year, attended medical school. Now residing in New York City, she is advancing toward an M.F.A. from

City College. Caitlin's work has appeared in both the *Bellevue Literary Review* and the *Texas Review*.

Paul Kvinta is a contributing editor at *Outside*, where he writes often about wildlife and the environment. He has also written for *National Geographic Adventure*, *GQ*, *The New York Times Magazine*, *Popular Science*, *Audubon*, and other publications. His work has won numerous awards and appeared in several anthologies. His story on human-elephant conflict in India won the Daniel Pearl Award, appeared in *The Best American Magazine Writing 2005*, and was a finalist for the National Magazine Award. He has been a Knight Journalism Fellow at Stanford University and a Templeton Journalism Fellow in Science and Religion at Cambridge University. He lives in Atlanta with his wife, daughter, two cats, and a bearded dragon.

John Lanchester was born in Hamburg in 1962. He worked as a football reporter, obituary writer, book editor, restaurant critic, and deputy editor of the *London Review of Books*, where he is a contributing editor. He is a regular contributor to *The New Yorker*. He has written four novels—*The Debt of Pleasure*, *Mr. Phillips*, *Fragrant Harbour*, and *Capital*—and two works of nonfiction: *Family Romance*, a memoir, and *Whoops! Why Everyone Owes Everyone and No One Can Pay*, about the global financial crisis. His books have won the Hawthornden Prize, the Whitbread First Novel Prize, the E. M. Forster Award, and the Premi Llibreter; been long-listed for the Booker Prize; and been translated into 25 languages.

Rachel Leven is an environment reporter for the Center for Public Integrity, a nonprofit, nonpartisan investigative news organization. She previously worked for Bloomberg Environment and *The Hill* newspaper.

J. B. MacKinnon is a freelance journalist and author, most recently of *The Once and Future World: Nature as It Was, as It Is, as It Could Be*. He lives in Vancouver, Canada.

Siddhartha Mukherjee, a cancer physician and researcher, is the author of *The Gene: An Intimate History* and *The Emperor of All Maladies: A Biography of Cancer*, the 2011 Pulitzer Prize winner for general nonfiction. He has published articles in *Nature*, the *New England Journal of Medicine*, the *New York Times*, and the *New Republic*.

Barack Obama is the 44th president of the United States; author of *Dreams from My Father: A Story of Race and Inheritance* and *The Audacity of Hope:*

Thoughts on Reclaiming the American Dream; and recipient of numerous awards, including the Nobel Peace Prize.

Elena Passarello is the author of two collections of essays, *Let Me Clear My Throat* and *Animals Strike Curious Poses. Animals Strike* was named a Notable Book of 2017 by the *New York Times, Publishers Weekly,* and the *Guardian.* The recipient of a 2015 Whiting Award in Nonfiction, she teaches in the M.F.A. program at Oregon State University.

David Roberts has been writing about climate change, energy, and the politics where they intersect for more than a dozen years now, first at the nonprofit environmental news site Grist.org and now at Vox.com. He lives in Seattle with his wife and two children.

Joshua Rothman is the archive editor for *The New Yorker.*

Kathryn Schulz is a staff writer for *The New Yorker* and the author of *Being Wrong: Adventures in the Margin of Error.* In 2016, she won the Pulitzer Prize for Feature Writing and a National Magazine Award for "The Really Big One," her story on the seismic risk in the Pacific Northwest.

Christopher Solomon (@chrisasolomon) is a contributing editor at *Outside* magazine. He also writes on the environment and the outdoors for *The New York Times Magazine, National Geographic,* and other publications. He lives in north-central Washington State, and more of his work lives at www.chrissolomon.net. His work also appears this year in *The Best American Travel Writing 2018.*

Kim Todd is the author of *Sparrow, Chrysalis: Maria Sibylla Merian and the Secrets of Metamorphosis,* and *Tinkering with Eden: A Natural History of Exotic Species in America.* Her work has won the PEN/Jerard Award and the Sigurd Olson Nature Writing Award. She is on the nonfiction faculty of the University of Minnesota's M.F.A. program and is at work on two projects: one on predators and one on the "girl stunt reporters" of the late nineteenth century.

Kayla Webley Adler is a senior editor at *Marie Claire,* where she writes and edits both the news section (called News Feed) and features. She has recently written stories on gender bias in medicine, an abortion fund hotline in Atlanta, and sex trafficking in North Dakota; edited a comprehensive how-to guide for women running for public office; and collaborated with *Esquire* on an ambitious project on workplace sexual harassment. Previously, Kayla worked as a staff writer at *Time* magazine, where she reported feature stories on innovations in education, international adoption, and school bullying, as well as contributing regularly to TIME.com and to the annual TIME 100 and Person of the Year special issues.

Ed Yong is a science journalist who reports for *The Atlantic* and is based in Washington, D.C. *I Contain Multitudes,* his first book, is a *New York Times* bestseller.

Other Notable Science and Nature Writing of 2017